ROUTLEDGE LIBRARY EDITIONS: HUMAN GEOGRAPHY

Volume 15

T0264923

MEDICAL GEOGRAPHY

MEDICAL GEOGRAPHY

Techniques and Field Studies

Edited by
N. D. MCGLASHAN

Routledge
Taylor & Francis Group

LONDON AND NEW YORK

First published in 1972 by Methuen & Co. Ltd

This edition first published in 2016
by Routledge
2 Park Square, Milton Park, Abingdon, Oxon OX14 4RN

and by Routledge
711 Third Avenue, New York, NY 10017

Routledge is an imprint of the Taylor & Francis Group, an informa business

© 1972 Methuen & Co. Ltd

British Library Cataloguing in Publication Data
A catalogue record for this book is available from the British Library

ISBN: 978-1-138-95340-6 (Set)
ISBN: 978-1-315-65887-2 (Set) (ebk)
ISBN: 978-1-138-99959-6 (Volume 15) (hbk)

Publisher's Note
The publisher has gone to great lengths to ensure the quality of this reprint but points out that some imperfections in the original copies may be apparent.

Disclaimer
The publisher has made every effort to trace copyright holders and would welcome correspondence from those they have been unable to trace.

Medical Geography

TECHNIQUES AND FIELD STUDIES

EDITED BY

N. D. McGlashan

LONDON

Methuen & Co Ltd

11 NEW FETTER LANE EC4

First published in Great Britain in 1972
by Methuen & Co. Ltd
11 New Fetter Lane, London EC4
© *1972 Methuen & Co. Ltd*
Printed in Great Britain by
Butler & Tanner Ltd
Frome and London

SBN 416 65400 2

Distributed in the U.S.A. by
HARPER & ROW PUBLISHERS, INC.
BARNES & NOBLE IMPORT DIVISION

Contents

* Reprinted from journal articles

* *Reprinted from journal articles*

List of Figures

Acknowledgements

The editor wishes to acknowledge with thanks permission of the following editors to reprint certain chapters which were first published in journals:

Chapter 4: Forster, F. (1966) *British Journal of Preventive and Social Medicine*, 20, 165–71; Chapter 6: McGlashan, N. D. (1968) *Central African Journal of Medicine*, 14, 249–52; Chapter 8: Prothero, R. M. (1967) The Sixth Melville J. Herskovits Memorial Lecture delivered under the auspices of the Program of African Studies, Northwestern University on 9 October 1967; Chapter 12: Fonaroff, L. S. (1968) *The West Indian Medical Journal* 17, 14–21; Chapter 15: McGlashan, N. D. (1967) *Tropical Geographical Medicine*, 19, 333–43; Chapter 19: Hunter, J. M. (1966) *The Geographical Review*, 56, 398–416; Chapter 21: Tinline, R. (1970) *Nature*, 227, 860–2.

The editor is also most grateful to the University of Tasmania for a grant to assist with drafting expenses and to Mrs. D. Stuetzel for most of the cartography and Miss J. Young for typing the manuscripts.

Andrew Learmonth has been a constant source of kind encouragement and experienced advice.

Neil McGlashan
Hobart
16.3.71

Part I
Introductory

1 Medical Geography:
an Introduction

N. D. McGlashan

Any collection of papers in a borderline discipline faces the difficulty of satisfying readers drawn from two specialities. Inevitably much must be said that is 'old hat' to one or other group if both are to comprehend the value of the partnership. Often both groups introspectively know their own position but, for others, have little regard. For this reason geographers coming to the obvious remarked on in these pages are asked for their forbearance. Similarly physicians are asked to overlook geographers' possibly naïve approach to medical problems.

This collection has been put together entirely by geographers in part to indicate the type of work currently being undertaken. We include no retrospect and the book contains only work undertaken in the last decade. Co-operation is invited of the medical profession in discussing their problems from a spatial viewpoint.

THE DEVELOPING RELATIONSHIP

The enormously rapid growth of both theoretical and practical knowledge is accepted in the world today as we accept, albeit with unease, the likely doubling of earth's human population before 1999. We also may expect a doubling of mankind's collective range of understanding. Assuming individual mental ability static, the situation is inherently bound to increase academic specialisation with an increase in sub-specialities within existing fields and also an increase in studies which straddle old inter-subject borderlines.

One such area is the common interest between geography and medicine which has only been *widely* recognised in the last twenty years. Broadly the field lies in local variations of those environmental conditions which are causatively related to human health or ill-health. This relationship was known in the fourth century B.C. to the Hippocratic school of healers. Their words provide so apt a description of the subject that medical geographers today often cite them to illustrate the long ancestry and, by implication, the reputable standing of their subject. In the following quotation three geographical variables are stressed: climate, site and occupation. The first two correspond clearly to physical geographical factors and the last to human social geography representing both economic effort, say, chimney sweeping or asbestos mining and recreational custom, say, athletic exertion or usage of tobacco.

> He who wishes to study the art of healing must first and foremost observe the seasons and the influence each and every one of them exercises . . . and further he shall take note of the warm and cold winds . . . so should he also consider the properties of the water . . . The healer shall thoroughly take the situation into consideration and also the soil, whether it is without trees and lacks water, or is well wooded and abundant with water, whether the place lies in a suffocatingly hot valley or is high and cool. Also the way of life which most pleases the inhabitants; whether they are given to wine, good living and effeminacy, or are lovers of bodily exercises, industrious, have good appetites and are sober.[1]

With so clear a directive to 'the healer' provided so long ago it is pertinent to inquire why medical geography, despite some significant nineteenth-century studies, is developing only recently. I believe that the answer lies very largely in technology. Descriptive types of geography were totally ill-equipped to undertake studies which depend upon quantitative and comparative geographical analysis. For instance observation 'of the seasons and their influence' depends upon the invention of the thermometer; 'the properties of the water' can only be geographically studied when chemical trace-element analytic techniques have been discovered. Spatial variations in diseasedness cannot be studied until basic demographic information about populations is available and until scientific standards of medical diagnosis and recording are set. Contributory information from many other natural sciences is needed to obtain the full and ideal Hippocratic data-base; climatology, pedology, botany are merely some amongst many.

Even when these ancillary and mensural techniques have been developed within their own sciences they are not all locally available everywhere. Many parts of the world, for instance, still lack soil survey information. Then, at a later stage, the sheer volume of natural scientific data becoming available threatened to swamp geographical analysis. Two more recent developments are tending to combat the analytic inadequacy. The development of quantitative methods of multi-factoral geographical appraisal has sprung largely from a similar bond between statistics and geography. The basis of the underlying concept hinges upon the recognition of groups of factors significantly co-extensive with areal patterns of disease or death. The methodological framework for these studies within geography descends from regional analysis where relationships are sought, in geographical space, between diverse factors. The culmination of the argument would be to reach a point where one may recognise causation; 'x' plus a little 'y' causes 'z'. Applied to medicine this is aetiology. Its importance lies in identifying previously unrecognised causative factors as a preliminary to removing or controlling factors harmful to man. Quantitative precision in handling diverse data is also of value in the confidence with which scientific results can be repeated and generalisations inferred.

The invention of the electronic computer, currently becoming more and more capable in terms of power, provides the analyst with equipment which not only removes from him the necessity of laborious calculation but allows him to conceptualise fields of analysis previously completely closed to human calculation by the span of life-times needed to work through the mathematics. Thus, and only in the last two decades, has it now become a practical feasibility to undertake work on the lines formulated by the Hippocratic school.

WHAT IS MEDICAL GEOGRAPHY?

Although some major technological impediments to work in the borderline field have been removed, there are still divergent opinions on the nature of the subject. That these varying views are held is surely entirely healthy, for they should lead to the production of useful studies in a wider area. At the same time the limitations imposed by one's particular professional training should give as clear an idea of what should not be attempted as of what can be done. For example, medical geography in the U.S.S.R. claims a far longer basis of research experience than is the case in the West and it is now an independent branch of science.[2] Emphasis is placed upon studies of multi-factorial disease complexes[3] and upon public health as a contribution to the productivity of labour and hence to the development of a national economy on socialist lines. This pragmatic aim is particularly evident to Soviet regional medico-geographical studies of under-developed zones in produce public health forecasts of regions intended in the future for economic development. Overall health is seen as a measure of the totality of health risks inherent in any particular environment[4] and, conversely, regions especially salubrious are designated suitable for such uses as sanatoria.

In the West, medical geography has been used rather as an aetiological research tool with emphasis on reducing mankind's load of suffering from ill-health. Freedom from centralised direction of research funds and hence individual choice of topic and scope has resulted in individual contributions often by geographers working in isolation in their chosen 'emerging speciality'.[5] Perhaps the specialisation lies in the type of data which an individual geographer most often uses. Constant study in particular groups of spatially varying data lead understandably to an increased awareness of their limitations and of ways of overcoming these. To this extent all geographers have a common bond in the types of questions they ask, the viewpoint they develop and the analytic methods they employ; sub-groups may be recognised by the phenomena studied; farm classification by the agricultural geographer, beach profiles by the geomorphologist or mortality patterns by the medical geographer. May has questioned whether there is 'such a thing as a distinctive method applicable to the systematic acquisition of knowledge in medical geography'.[6] The answer hinges most crucially on the phenomena selected for study by normal geographers' methods. May defines geogens as factors of environment which have, or are thought to have, a relationship with disease causation. Alexander and Armstrong[7] place the same emphasis on subjective judgement when they emphasise study of factors 'believed to be significant'. Thus every individual worker has choice of what factors he *believes* may be relevant and that he will include in his field of study. Since most diseases are known to be multi-factorial in aetiology the relevant factors for study rapidly approach infinity and some form of selection is necessary.

Another aspect which has received emphasis is that health too needs to be studied in its spatial variations, in which the concept of 'degrees of health'[8] makes up a continuum from the ideal of absolute physical, mental and social well-being through variations of diseasedness to death. Disease is thus looked upon as a lack of adjustment between our living cells and the challenges consequent upon their existence within an environment as normally studied by geographers as well as within the micro-environment of parasites, viruses, harmful chemicals and so on. Clearly this total environment is also dynamic: man builds dams increasing the *liebensraum* of bilharzia-carrying snails and malarial mosquitoes. Man alters his own diet and environmental stresses leading to 'bottle diarrhoea' in Uganda[9] and increased incidence of appendicitis in central Africa.[10]

To narrow down a topic for study in this vast field must be chiefly a matter of individual interest and experience, but some part depends admittedly also upon luck. Even in an analysis of many factors the one *critical* factor may be omitted, either from a real lack of that information or from the need to find a working limit for the bounds of the study. For this reason alone many medical geographical studies come up with inconclusive answers – 'other factors may (truly) need to be considered'. The level of explanation by geographical analysis alone may not always progress beyond association-in-space. Nonetheless geographers today are increasingly aiming research towards providing positive and scientific explanation.

MEDICAL GEOGRAPHIC DATA

The difficulties in handling epidemiological data[11] may be looked at briefly from the geographer's point of view in four groups.

The strictly medical information upon health, ill-health or death is not usually gathered in the neatly compartmentalised way in which we would like it. Even in the most straightforward case, certification of cause of death, what of the cancer patient killed in a car crash? Should we look beyond the primary and seek secondary or tertiary potential causes? But these may not have been diagnosed; and, if diagnosed, may not have been fully recorded since only an atypical group of corpses, possibly of forensic or medical problematic interest, reach autopsy. There is the question of medical 'fashions' in diagnostic terminology, the special interest or skills of the physician and his consideration for surviving relatives, all of which may influence the cause of death as recorded.

Dealing with morbidity, some examples of difficulties with data include these. Confidence may be breached by allowing access to patients' particulars; some conditions may be socially inadmissible, extra-marital pregnancy or psychiatric disturbance for instance. Diagnosis may alter after being first recorded; the patient may suffer from several conditions and be recorded only under the most treatable one. Comparability of diagnostic standards as between different practitioners; differences of patient recognition of sickness and degrees of effort put into seeking treatment; cost of treatment and, in some places, religious flavour with treatment; all these need to be studied and allowance made.

The second main group of difficulties with the data may be classed as medical statistical; the availability of adequate runs of comprehensive figures of vital statistics – still woefully lacking in much of the world. Measures of standardising those figures one wishes to use and a scientific awareness of how they can be utilised; alternatively, in undeveloped lands, ways in which some usefulness can be derived even from partial data;[12] these are part of the fascination of medical geography.

Thirdly there is the question of the locational specificity of the data. For instance, medical data are often collected from hospital statistics which may not include for each patient a home address. Data may refer to administrative divisions, and home and work places lie in different divisions. Indeed occupational risks to disease may arise from previous or present employment, from journey to work or any one of various home places or social contacts. All need locating. Census data may utilise different areal units to those referring to morbidity or mortality information. Then for base maps it may be more suitable to use a demographic base instead of a normal map,[13] whereas, in some places, even the most basic topographical map may not exist. The handling of problems of this nature are one's stock-in-trade.

Fourthly lies the choice of other data to be considered. As already explained, this is often an intensely subjective matter. Vast stores of information are held by government agencies in most countries, but these will not always be accessible or easily assimilable. Such information, say variations of staple plant yield for example, may not exist or may first have to be processed and mapped before use. Where data are readily available, mapped and published, are they up-to-date and the best available? Where data seem to be non-existent (as they often are concerning local and intimate details of custom), the medical geographer may face all the usual problems of devising a questionnaire suited to his purpose and geared to the understanding both of his staff and of the respondents. Acland showed initiative in studying Cholera in Oxford in 1856[14] by obtaining from an engineer a map specially for his purpose with contours given at 5 ft. intervals.

DATA PRESENTATION

It has been claimed[15] that the Golden Age of medical cartography was between 1835 and 1855. It is certainly true that the map is still one of the main tools of geographical interpretation today. Petermann, writing in 1852,[16] well explains the advantages of maps for medical purposes.

> The object, therefore, in constructing Cholera Maps is to obtain a view of the Geographical extent of the ravages of this disease, and to discover the local conditions that might influence its progress and its degree of fatality. For such a purpose, Geographical delineation is of the utmost value, and even indispensable; for while the symbols of the masses of statistical data in figures, however clearly they might be arranged in Systematic Tables, present but a uniform appearance, the same data, embodied in a Map, will convey at once, the relative bearing and proportion of the single data together with their position, extent, and distance, and thus, a Map will make visible to the eye the development and nature of any phenomenon in regard to its geographical distribution.

These maps may be prepared with wide variations both of care and of accuracy. Since the map may be read, used, acted upon by other professions, it is of utmost importance that a map's limitations and total dependence on its data base be also explicit upon its face. It is easy to convey an impression of totally spurious reliability.

Maps too can (obviously) be prepared with varying amounts of detail, from simple letters or figures at the site of the report to the computerised interpolated isopleth of a mortality value based on full age/sex standardisation. Medical geographical mapping will be further considered below as one means of spatially defining medical data.

Geographical skill in selecting appropriate means of display does not necessarily end with the map. Diagrams and histograms of various sorts have usefulness and, especially, graph-type displays are particularly suited to problems involving the associated changes of value of two parameters, say a disease and a measure of recreational time.

PUBLIC HEALTH ADMINISTRATION

This collection starts with a group of general or introductory papers in Part I. Following this discussion of our subject matter, Learmonth reviews recent medical work in the borderline field in Chapter 2 and shows that geographers are not alone in these approaches to chiefly epidemiological studies. Then Copperthwaite compares, with examples from Yugoslav Macedonia, the results obtained in one rare case where both mortality and morbidity statistics are available. Forster, a geographer at the Usher Institute, contributes Chapter 4 which explains the mechanics of preparing a base map proportional to size of population, dividing the population into groups at risk by age and sex. Finally in the introductory group, Armstrong illustrates the role to be played by computer line print-outs for speedy, but fairly rough, map-work.[17]

Part II considers matters concerned with distribution of medical facilities. Amongst the classics in this field but too long for inclusion here is Godlund's[18] study in 1961, which considered population forecasts and transport facilities in Sweden in relation to the vast capital expenditure then contemplated in up-grading the facilities of certain selected hospitals to make them centres for region-wide medical networks. He used a measure ('isochrones') of equal travel time about the possible centres to assess those best placed for costly development.

Cook and Walker [19] published a study in 1967 of regional variations in dental care in the United Kingdom using an index of general dental practitioners per ten thousand of population and repeated after ten years. When compared the results showed a very marked regional imbalance in services and possible measures were discussed towards alleviating the inequitable situation.

A growing number of health authorities across the world are becoming aware of the potentialities of this type of study which is of increasing importance with the growing capital expenditure upon ever more costly hospitals and medical equipment. Even in undeveloped countries there is a parallel need to spend each kwacha, escudo or what-have-you of the health vote as rationally as possible.

This collection includes two short papers (Chapters 6 and 7) from undeveloped countries which consider ground- and air-based medical services in Malawi and Zambia respectively. These are followed by an account by Prothero of part of his well-known work with the World Health Organisation in Sudanic Africa. Here he discusses the problems of combating disease in the peculiarly difficult conditions of the pastoral folk of Somalia. Concluding Part II and as a contrasted study from an overdeveloped country, Armstrong contributes Chapter 9 on the role of geographical skills in health surveillance work in the United States.

Other matters of public health importance are maps of states of general nutrition, parasites and disease vectors as well as changing morbidity patterns through time. These give opportunity to supervise a disease situation with regard to appropriate quarantine or control measures. To have the necessary information in a geographical form on computer call may become an important role of medical geography. The form of output might look remotely like a synoptic weather chart, 'an epidemic "high" of unidentified fever is moving slowly north-east across Spain and is expected to reach the Channel ports of France and England within a week or ten days'.

SPATIAL DEFINITION

It is often simplistically said that the geographer's task is to answer the question 'where?'. That is to localise, and usually portray, data which relate to variation in space. Part III presents a group of papers which concentrate upon problems of spatial definition of disease. Chapter 10, a review of atlases of disease, drawing upon Learmonth's long association with this field of work, portrays many widely differing aims and standards of presentation usefully dividing world-scale from developed from undeveloped regions' information.

In his task, the geographer has available to him, broadly speaking, four types of mapping. Firstly he may choose to show every occurrence – death, disease, cabbage patch or stagnant pool – by some form of dot distribution. The known difficulties about the placing of a dot to represent multiple occurrences need care. Such a map has many uses particularly in the administration of medicine and in cluster analysis. However it often has the disadvantage of showing only a 'pale shadow' of the total (statistical) population distribution. To avoid this criticism, the number of cases within an area can secondly be related to the population to produce a 'rate'. If the rate refers to disease, it is usually expressed as so many per hundred thousand persons. If the frequency of occurrence and the detailed data permit, subdivisions by age and sex can be separately portrayed. The technique is illustrated by the transformation of a *dot map* of blindness cases in one province of (then) Northern Rhodesia into a *rate map* in Chapter 11. This is followed by Fonaroff's neat display of the decline of malaria in Trinidad as shown by spleen-rate data (Chapter 12).

The resulting choropleth maps may then be re-mapped in a third manner to treat the area as a continuous surface and isolines of disease or death – isomorbs or isomorts – may be constructed as contrived cartographic devices. Learmonth's paper on atlases of disease (Chapter 10) illustrates this form of mapping from his own work in Australia.

The fourth type of map relevant to the portrayal of disease and death is a probability map which sets out to recognise occurrences which are part of a *non-random* spatial pattern. Chance fluctuations are, of course, unimportant and distributions may be tested by chi-squared, Poisson or other statistical tests, applied to each area. One would expect greater significance usually to attach to areas with larger numbers of cases. These larger numbers can be obtained either by taking a run of several years' figures or by increasing the size of the unit of area in use.[20] Temporal aggregation may, however, mask gradual secular change and areal aggregation conflicts with the geographical requirement to narrow down the spatial exactitude of the map. Part III of this book concludes with two papers exemplifying significance mapping. For assessment of non-random occurrence of leukaemia in England and Wales, White uses probability maps (in Chapter 13) based on aggregation of varying time periods. In Chapter 14, which utilises data on white deaths from male stomach cancer in South Africa, the emphasis is upon mapping as a means of bringing out significance by differing criteria; both by the 'tails' of the normal statistical distribution and, alternatively, against the Poisson expectation.

Such a map of significantly high or low rates of disease or death may be looked upon as a direct stimulus to aetiological reasoning. Other environmental 'indicators' may at once be apparent and the screening of potentially culpable factors is a step towards the search for causes.

ASSOCIATIVE OCCURRENCE

The comparison of environmental factors for an area can also be put upon an objective footing by the use of techniques of statistical evaluation to assess just how alike distributions are and at what level such a similarity might not have occurred by chance. Part IV of this book is concerned with this spatial correlation and starts (Chapter 15) with what may be considered a cautionary tale about the interpretation of geographical results in medicine.

Comparison may be in pairs severally – a disease with a socio-economic factor – or it may, with computer facilities, run into multiple regression. The type of regression selected will depend upon the aetiological meaning to be assessed. Girt and Dever, both working in developed countries, illustrate in Chapters 16 and 17 the comparison of a pattern of disease, chronic bronchitis and leukaemia respectively, with a number of social and housing variables. Girt emphasises the role of past 'risk' environments and Dever shows the varying results obtainable as the scale of focus upon a problem is varied.

This leads to the vital problem of recognising causation. The geographer studying population generalisations will rarely, if ever, prove a causative relationship. He may well draw attention to previously unrecognised inter-relatedness between variables. Clearly, for conditions whose aetiology is already known, these associative steps are unnecessary. This method of approach has its greatest potential in the study of any disease whose cause still awaits elucidation. For example, a study in Chapter 18 of alcoholic drinks as carcinogens represents correlational work with a necessarily less strictly quantitative basis originating in central Africa.

Interesting examples of this type of approach not included here for reasons of space, are provided by Sakamoto[21] with her studies of seasonal variations of disease incidence amongst racial groups in the United States. Another environmental factor, which has been studied by Warren,[22] is soil trace-elements in relation to goitre and multiple sclerosis. As Armstrong[23] has pointed out (in another context) the causative pathway from soil to man is a long and complex one, liable to misunderstanding at many points, especially in a world where few of us today eat predominantly or even significantly of the products of local gardens. Davidson's[24] team, working upon high diabetes incidence where cassava (manioc) is the staple diet, found a similar difficulty in *proving* a physiological reason for this apparent association in space.

In brief, the medical geographer's tasks are to prepare and collate disease data and to map them to show where a certain condition is rife (or absent); to apply objective statistical tests to these distributions to assess whether or not the pattern is likely to have occurred by chance; to measure the degree of co-extensiveness between disease and other spatially varying factors; and then to apply tests to decide whether any spatial associations he has shown could be causative. Many such inquiries will come to nothing, but the initiation of even a single new hypothesis will be of value to medicine.

DISEASE DIFFUSION

Another variable of vital concern to the geographer is time – the changing distributions of phenomena through time. A very simple example is the finding that some seasonal peaks of disease incidence are becoming less through time.[25] Another aspect is the cumulative effect of some disease-causing mechanisms through time – a case in point is the smoking/lung cancer relationship.

Rather more sophisticated in approach are studies of disease spread through time *and* space. For the geographer, epidemic waves are similar as a topic to other studies of innovation diffusion in which, to borrow sociological language, one can identify innovators, adopters and immunes. An outward spread from a centre can be postulated, like ripples from a stone thrown into a pond. The advance of an 'epidemic wave' can be mathematically modelled and a series of possible barrier or enhancement effects can be tested on the model until it is shown to conform to reality. The ultimate purpose, of course, is to advise public health authorities regarding appropriate control measures even when the mechanisms causing the speed of spread are not fully understood at a medical level. In Part V of this collection Hunter demonstrates an ecological approach to the transmission cycle of an infectious disease. This is followed by Brownlea's and Tinline's papers, which illustrate the ways in which geographers look at specific problems and both show ways likely to lead to considerable saving of public expenditure. In Australia infectious hepatitis and in Britain foot-and-mouth disease constitute financial disasters.

Work upon disease diffusion is very close to modern quantitative geography with its emphasis upon flows and networks.[26] The final paper, Haggett on measles' spread, shows that geographers in this field of study may contribute indirectly to reducing human suffering from disease by furthering our understanding of causative processes.

THE I.G.U. COMMISSION ON MEDICAL GEOGRAPHY

In all these tasks geographers follow the late Sir Dudley Stamp who referred to medical geography as a 'tool for research'.[27] With the increasing tendency to modern quantitative analysis in geography, however, it is a tool likely to prove ever more useful. The role of the medical geographer is to make available to medicine the skills of geography but in no way to usurp the functions of medical men. Unfortunately Sir Dudley did not live to see the subject recognised at New Delhi in December 1968 as a full Standing Commission of the International Geographical Union. Although the call for Commission status was supported by most of the member nations, it is the developing countries which probably have more to gain than the developed ones. In terms both of disease distribution and of environmental factors Birmingham and Sheffield may seem fairly similar. There are great differences, however, between Barotseland and the Zambian Copperbelt. This point leads into another. Anglo-America and North-West Europe have today an almost monolithic social culture. Parts of Africa and Asia vary very greatly from each other. 'In parts of Africa the disease picture resembles that of medieval Europe for the individual may suffer from five or six diseases.'[28] In other areas, rising incomes and literacy standards are bringing Africans closer to the public health patterns of Europe[29] and the onrush of development is rapidly doing away with the heterogeneity of the past. This gives real urgency to studies based upon areal differentiation before those differences disappear. In turn this urgency justifies studies in medical geography being attempted without delay. It justifies the acceptance of today's low standards of accuracy and it rejects delay for improvement of diagnoses and statistical data, although these may be possible in the future.

There remains the question of who is to undertake this type of work?

Surely the question itself is barely relevant, for as Fonaroff and Banta put it 'disciplinary boundaries will hopefully remain hazy'.[30] Collaborative effort as co-members of an inter-disciplinary team is likely to yield best results and even the disciplines represented will vary with the individual problem which requires solution. Most often medical geographers will work with medically qualified men, but it may be in botany or in biochemistry that a particular piece of interpretative scientific search will lie.

Here geographers recognise the need for caution. Medical geography is a tool and but rarely an end in itself. It is the application of geographical methods and skills to medical problems. One may consider 'geographical *evidence* on medical hypotheses'.[31] It would be improper to claim that geography provides *proof*. Geography inevitably must generalise about an area and, in doing so, its hypothetical relationships are also generalisations. The confirmation needed for such hypotheses will lie with a discipline which, rather than the group, studies each individual case.

REFERENCES

1 HENSCHEN, F. (1966) *The History of Diseases*, London, and Hippocrates transl. Jones, W. H. S. 1923, Cambridge.

2 BYAKOV, V. P. *et al.* (1962) *Soviet Geography: Accomplishments and Tasks.* American Geographical Society.

3 IGNATYEV, YE. I. (1969) *Key Problems in Mapping in the Field of Medical Geography.* Dept. of Army transl. No. J-6384 Washington.

4 SHOSHIN, A. A. (1968) *Fundamentals of Medical Geography.* Dept. of Army transl. No. J-5264 Washington.

5 ARMSTRONG, R. W. (1965) *Int. Pathology*, **6,** 61.

6 MAY, J. M. (1954) 'Medical geography'. In P. E. James and C. F. Jones (eds) *American Geography: Inventory and Prospect.* Syracuse.

7 ALEXANDER, C. S. and ARMSTRONG, R. W. (1966) *Illinois Veterinarian,* **10,** 19.

8 BANTA, J. E. and FONAROFF, L. S. (1969) *Prof. Geogr.*, **21,** 87.

9 LANGLANDS, B. W. (1968) *The Uganda Atlas of Disease Distribution.* Kampala.

10 MCGLASHAN, N. D. (1968) *The Distribution of Certain Diseases in Central Africa.* Unpublished London Ph.D. thesis.

11 CASE, R. A. M. and DAVIES, J. M. (1964) *J. Inst. Statisticians,* **14,** 89.

12 COOK, PAULA J. and BURKITT, D. (1970) *An Epidemiological Study of Seven Malignant Tumours in East Africa.* Medical Research Council, London.

13 HUNTER, J. M., and YOUNG, J. C. (1968), *Prof. Geogr.* **20,** 402.

14 ACLAND, H. W. (1856) *Memoir on the Cholera at Oxford,* 1854.

15 GILBERT, E. W. (1958) *Geogrl. J.*, **124,** 172.

16 PETERMANN, A. (1852) quoted by Gilbert *op. cit.*

17 See also MCGLASHAN, N. D. and BOND, D. H. (1967) *Die Erde*, **98,** 292.
MCGLASHAN, N. D. and BOND, D. H. (1970) *Canadian Geographer*, **14,** 243.

18 GODLUND, S. (1961) *Population, Regional Hospitals, Transport Facilities and Regions*. Lund Studies in Geography No. B.21.

19 COOK, PAULA J. and WALKER, R. O. (1967) *Brit. Dent. J.*, **122,** 441, 494 and 551.

20 MCGLASHAN, N. D. (1966) *Int. Pathology*, **7,** 81.

21 SAKAMOTO, M. M. and KATAYAMA, K. (1966) *Papers in Meteorology and Physics*, **17,** 279.

22 WARREN, H. V. (1964) *Geogrl. J.*, **130,** 525.
WARREN, H. V. (1965) *Science*, **148,** 534.

23 ARMSTRONG, R. W. (1962) *Prof. Geogr.*, **14,** 1.
ARMSTRONG, R. W. (1964) *Acta Agric. Scandinavica*, **14,** 65.
ARMSTRONG, R. W. (1964) *Am. J. publ. Hlth*, **54,** 1536.

24 DAVIDSON, J. C. *et al.* (1969) *Medical Proceedings (Johannesburg)*, **15,** 426.

25 SAKAMOTO, M. M. and KATAYAMA, K. (1967) *Papers in Meteorology and Physics*, **18,** 209.

26 HAGGETT, P. and CHORLEY, R. J. (1969) *Network Analysis in Geography*. London.

27 STAMP, L. D. (1964) *The Geography of Life and Death*. London.

28 BANKS, A. L. (1967) *Lancet*, i, 749.

29 WALKER, A. R. P. (1964) *S. Afr. med. J.*, **38,** 255.

30 BANTA, J. E. and FONAROFF, L. S. (1969) *op. cit.*

31 MCGLASHAN, N. D. (1967) *Trop. geogr. Med.*, **19,** 333.

2 Medicine and Medical Geography

A. T. A. Learmonth

INTRODUCTION

The Editor has decided to collect papers about medical geography by professional geographers. This essay, with his concurrence, seeks to redress the balance, to some extent, by reviewing some relevant medical literature. Of course there is a personal bias as the writer looks over medical writings and writers whose influence has been important over the twenty years of his own serious commitment to medical geography as a field of study. A more serious limitation is that this paper explicitly excludes the great weight of contributions from biologists, veterinarians and statisticians which must equally justify geographical assessment. Meantime, however, it is worth expressing an opinion that the separate streams of research are converging: the waters are beginning to intermingle, and soon we may see a single broad stream of research in medical geography even though fed by headstreams in different disciplines. Of this there is more in the conclusion to this paper, and there is also a separate discussion of atlases of disease on pp. 134–52, some mainly designed by medical men and some by geographers.

What criteria can one adopt in seeking to sift out material relevant for a short essay from the huge – and humbling! – flow of medical literature? The main criteria used here are:

(1) The use by the medical author of maps or other indicators of area distribution patterns as serious tools of analysis, and not as mere illustrations. (Perhaps I should add that I do not imply that all geographers invariably use maps in the more desirable of these two ways.)

(2) Sheer suggestiveness to the trained geographer, the throwing of fresh light on the causes underlying areal distribution patterns, flows or trends. The latter criterion is admittedly not clear-cut: subjective elements are involved. But the eclectic flow into geography is important and fruitful enough to justify this. There may be something of a paradox here. Some of the most suggestive and exciting contributions from medical men overlap very strongly with an ecological viewpoint – mainly with the ecology of the biologist rather than of

the sociologist, and in these the stimulus is often greater in proportion to the waning of the milleniar anthropocentric basis of Hippocratic medicine. The less a worker approaches as a doctor, the more is he likely to recognise the streams convergent from sister disciplines – but of course within medicine itself there is often a marked contrast between the approach of the practitioner and the medical scientist.

(3) For reasons only of space, the date of work will also be used as a limiting criterion.

MEDICAL CONTRIBUTIONS TO MEDICAL GEOGRAPHY SINCE 1945

Here, for convenience – but by no means in watertight compartments – contributions are grouped into headings: *medical ecology*, including mainly studies founded on an intimate, local scale; and of the *broader studies of areal distribution patterns* which we may think of as more geographical in some sense, problem-oriented contributions are discussed, followed by those which are *regional-synthetic* in intent.

MEDICAL ECOLOGY

Geographers should never lose sight of the contribution of Dr. J. M. May to medical geography. It is massive, as even a selective bibliography shows, but if one is forced to select the contributions most seminal at least up to the present, there is a paper addressed directly to us.[1] Disease does have marked areal distribution patterns, and therefore a geography, but he urged that if we study it we must look to the number of factors involved in a particular disease-complex, and to conceive that the geography of a disease might depend on one, two , three of even four factors, each with its own geography.

An example of a four-factor complex is the tsutsugamushi disease (also called 'scrub typhus'). A mite of the subfamily *Trombiculinæ*, a rodent (*Microtus montevelloi* or a related species), a rickettsia (*Rickettsia orientalis*), and man constitute the complex. Before it becomes a disease of man, it exists on the soil as a partial pathological complex, since the adult mite feeds on the juice of the leaves of young grass that has grown in patches of cleared jungle abandoned after temporary cultivation. The larval mite may be brought into the area by birds or by rats in search of food where man has been living. The larval mite sucks the rickettsia from the blood of the rat. It survives through the nymphal and adult stages into the next larval generation. Man enters the diseased area and is bitten. The symptoms he develops are called the 'tsutsugamushi disease'.

This example illustrates the importance of silent zones of disease – zones where all the elements of disease are present except man. If man steps into

these silent zones without immunity and in a receptive condition, the phenomenon of disease occurs. These zones are widespread.

No geographer approaching this field of study should omit to read at
least the first thirty-four pages of May's *The Ecology of Human Disease*.[2]
This includes chapters on the nature of disease ('From the ecological point
of view, disease is very simply that alteration of living tissues which jeopard-
izes their survival in their environment'); on stimuli – inorganic, organic
and socio-cultural; on responses to environmental challenges; on the terrain; on races or populations; and on cultural factors. This last chapter
includes a classic passage (pp. 30–2), reproduced here though it might well
come rather under the later section on regional synthesis:

In Viet Nam as elsewhere the earth, the waters, the winds, and the heat had
organized themselves in ways which had resulted in today's relief, hydrography, and climate. In our times this consists of three concentric terraces
presiding over a flat lowland bordered by the sea in much the same way as the
balcony, the dress circle, and the orchestra surround the stage. At the top a
few tribes of hardy mountaineers dwell, very sparse, not afraid of fighting big
game and living chiefly by hunting and gathering. They have their own
diseases, rarely seen by Western eyes because they very seldom come down to
the plain where Western culture has begun to penetrate. In the hilly region
(the dress circle) there are more villages, although still quite sparse, because
only ten per cent of the land can be put to agricultural use. In this middle
region no big game, but an abundance of *Anopheles minimus*, a malaria bearing
mosquito, which thrives in the clear running water of the hill streams.

In the lowlands: a teeming population, grouped in compact villages where the
density in places reaches 1,500 per square mile. In this lowland the Red River
is the undoubted imperial master of the country. The river carries fertile
alluvial topsoil with which it covers the countryside but also danger in the form
of devastating floods which occur every year precisely at the time when the
second rice crop is ready for harvest. It also shifts its course from one bed to
another. Dense population occupy the temporarily abandoned beds, gambling
that one, two, or perhaps ten or fifteen harvests of rice can be grown before
the neutral river sweeps away crops, houses, and inhabitants. In certain places
the effect of the tide makes itself felt and there are large pools of salty water,
which has a nefarious influence on the crops of the hardy farmers who try to
put the land to good use. Thus for many reasons dams have been built ever
since recorded history with various degrees of technological skill for the triple
purpose of protection, irrigation and drainage.

The impact of these water problems on the local life is terrific. Without the
dams one crop out of two would probably be lost, and widespread starvation
would occur. To keep the dams in working order, swarms of laborers have
to be recruited, sometimes in a hurry, thus taking them away from the exacting
labors of the rice fields, and lowering the yield of the crop. It is a matter of
computing which way the least is lost, either making a smaller harvest as safe

as possible or taking a chance that the river shall not rise to the point of destroying a larger harvest. At one place or another, year after year, populations crowd together in very large numbers, away from their homes, bare legs in the mud, working to the point of exhaustion, plugging, digging, and plugging again to stop the infiltration of water, helped in more recent years by an organized distribution of rations. Once the work is completed, they return to their villages, sometimes carrying with them the seeds of cholera, collected from other workers who had brought them from their own contaminated hamlets. This need for a shifting labor force, however, is not the only reason for mass migration with their pathological and genetic consequences within the delta.

The harvest time is not the same in the north as in the south, in the west as in the east. Armies of field workers concentrate in the regions where the harvest is earliest and gradually move to the places where it occurs latest. You can see them moving on foot on the narrow paths between the rice fields, covered with a waterproof raincoat of tightly sewn banana leaves, carrying their sickles, and sometimes a bamboo pole and two baskets, with which they do a little trading on the side.

Because the land is scarce and also because until the French established the *Pax Gallica* by the end of the 19th century, people had to protect themselves against each other; the villages are compact, easy to defend, if possible surrounded by moats or thickly fenced in by a bamboo hedge.

The most favored inhabitants have their homes near the rim of the village and have a sort of jetty which juts out onto the pond. This is convenient. It is a sort of bathroom attached to the house, where the family can take a shower by ladling out the chocolate colored water. It also provides a kitchen sink to wash the rice, the dishes, and the laundry. It is also convenient for the raising of fish and water cress used to feed the pigs, and in the lean years after the pigs have been eaten to feed the men. However, the water itself is not drunk, except on rare occasions. The Vietnamese of the delta likes to drink rain water which he collects in cisterns, sometimes in jars and bottles, which he sells as medicine after it has been held many years.

If the climate, land forms, and customs govern the layout of the villages, they also govern the layout of the houses. The material of which their dwellings are built is, of course, the material that is available – its architectural potentialities govern the type of house. Lumber is scarce in the delta but abundant in the mountains; mud and straw are common in the delta but scarce in the mountains. We shall have to remember this when we try to understand the malaria pattern. In most dwellings there is a central yard and in the back the house proper and the altar of the ancestors. On each side of the yard in separate buildings, are the storeroom where the rice is kept and the pigsty. The house itself is covered by a roof, rarely of tiles, but usually of rice straw thatch. The floor is beaten earth and the walls are dried mud from the rice fields, filling in the spaces of the woodwork.

Now that we have seen where these people live, let us see what they eat. Rice, of course, and polished rice at that. Of course production of enough rice to support the population is the problem. We will not discuss here the fantastic edifice of mortgages and debts which rises above the fraction of an acre of land on which the family life is built. Nor shall we describe the land tenure laws and customs that have resulted in the reduction of the size of property through the years to insignificant proportions. Suffice it to say that both factors have a definite significance in terms of scarcity and starvation, hence a direct influence on the pathological picture of the country.

The problem of fertilizing these fields is vital. The farmers of the delta are well aware of the value of human fertilizer. In the villages human night soil is carefully collected, and in some places where there is a surplus, a brisk trade is initiated with villages of lower yield. In the evenings after their day's work is finished women gather the stuff and start on their endless trips balancing it on their shoulders in baskets at both ends of their bamboo poles. Urine is also collected and used for the most promising cultures. Animal fertilizers from pigs and buffaloes is also used as well as end products of fish which have been allowed to rot to produce a local spice called *nuoc mam*.

During two thirds of the year the delta is practically under water. Hence every farmer is a fisherman. He sends his children after snails, crayfish, and crabs as soon as they can walk, and these are a welcome addition to the daily diet. In these private ponds that surround the villages, fish is developed as a local industry. A few specialists get the best species (those which do not feed on other fish) from the banks of the Red River and bring them back to these ponds for breeding. They are skilled in the diagnosis of the best places where fish can be caught and they have developed a number of tricks. They watch the clouds, which indicate the time and place where fishing is going to be most rewarding. They watch the twigs floating on the river, or the color of the foam, which gives a clue as to the origin of this particular mass of water and hence of the fish that can be found in it. They send their buffaloes in the narrow streams to shake up the mud in the middle, forcing the fish to swim near the banks.

In certain areas of the delta the water people concentrate in floating villages, making a living by fishing or by water transportation. They pay dues to the nearest village for the privilege of burying their dead and sometimes of having a place of worship and a town hall. But in most of the 100-odd such villages these civic and religious activities occur in a floating building.

All this is chiefly fresh water fishing. The offshore waters are too shallow, the sand banks extend too far into the China Sea to allow for the development of sea fisheries, and out of a population of several millions, only very few fishermen try the salt water for its product. This explains why besides rice and yams, it is chiefly crayfish, snails and fresh-water fish that compose the menu of the people. Of course, there are lean years and fat years, but in most cases the average consumption of food does not exceed 1,800 calories at best and is low in animal protein.

Thus the people who live in the delta of the Red River survive. They have acquired a culture which has so far allowed them to adapt to the environment in which their ancestors settled. If this environment has fostered a certain type of life upon human beings, it has done the same to animals and plants. Situations have been created in this fight for adaption which have a direct bearing for better or for worse upon the disease pattern, frequently by promoting or destroying the effectiveness of a vector or of an intermediate host.

It can be summed up thus: from the waters the people get their food, also their cholera, their dysenteries, their typhoid fevers, their malaria; from the earth they get their hookworm; from the crowded villages they get their tuberculosis and their yaws; from the type of housing they have been forced to adopt they get their plague and typhus; and from the food which earth, temperature, and rain produce, their protein deficiencies, their beri beri.

Scrub typhus, summarised by May above as a 'four-factor complex', was studied intensively by Audy. This writer, probably less well known to geographers, is perhaps unexcelled for the clarity with which the intimate ecological scale of work is linked with the broader areal distribution patterns of main concern to the geographers.

As long ago as 1949 he demonstrated the natural focus of the scrub typhus cycle among rodents and trombiculid mites, along the margins between grassy flood-plains and tropical and sub-tropical forest in Asia and Australasia.[3] The lack of anthropocentric bias is remarkable, not least for a study begun during a crisis of casualties from disease during the Burma campaign in 1943-6. He then linked this cycle between parasites, pathogen and alternate hosts with man-made foci sufficiently similar to the natural ones to substitute for them in continuing the cycle, though often with different rodent species – the base of hedges, the margins of rice fields and the like, in the countryside and even in cities. From these ecological bases he went on to thorough geographical analyses of the whole Burma region. Happily a retrospective account of this work, relaxed and remarkably appealing in its breadth of culture, has become available through publication of Audy's Heath Clark lectures.[4] He has continued to provide us with stimulating ideas on all scales up to the world scale, and here I cite two valuable viewpoints on medical ecology and medical geography which may be accessible to many readers[5] and quote two passages from his book.

What, then, is this 'ecological slant'? And indeed, what is ecology, about which we hear so much but seem to know so little? There have been some recent reviews of this subject, all agreeing that the essence of ecology is the study of entire systems as wholes. A system in this definition is an assemblage of interacting and inter-dependent component parts, comprising living and non-living components and all media manipulated by them or through which information is exchanged between them. A system at this level is known as an ecosystem, to distinguish it from such a system as an individual organism. We may therefore conceive of a series of studies, each concerned with a higher level of organization than the one before.

'Environment' is a term belonging to fragmentary ecological studies, but inapplicable to studies of ecosystems in which a house or nest, a rat and its parasites, an automobile and its exhaust, the air and the water and the plants and the soil are all components in a system. 'Environment' is a concept which appears only when one component is artificially isolated for study – temporarily, we would hope.

Meantime in U.S.S.R. arthropod-borne viruses in particular were troublesome enough in the expansion of settlement in Siberia and Soviet

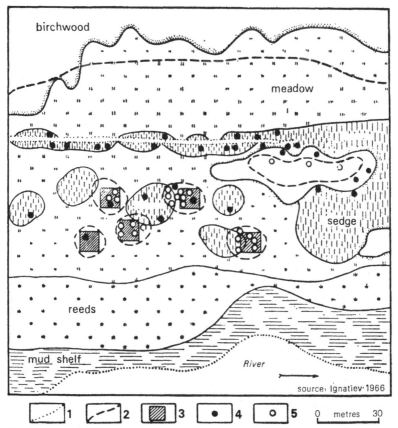

1 Water level 1950. 2 Water level 1951. 3 Foundations of houses. 4 Rat nests 1950. 5 Rat nests 1951.

FIGURE 2.1 Houses, rodent burrows and particular flood levels on the River Iskra, 1950, 1951.
(From: Fig. 5, p. 98 in Kucheruk, V. V. (1965) 'Problems of palaeogenesis of matural pest foci with reference to the history of the rodent fauna'. In A. N. Form-osov (ed.) *Fauna and Ecology of the Rodents*. Moscow Society of Naturalists, Proceedings of the Study of Flora and Fauna of the U.S.S.R., Moscow.)

Central Asia to cause a major research effort which evolved on somewhat parallel lines.[6] Perhaps the best way to indicate lines of work to geographers is by maps of two different scales (Fig. 2.1 and Fig. 10.7 in the essay on 'Atlases in Medical Geography, p. 146).

BROADER GEOGRAPHICAL ANALYSES

The work from U.S.S.R. makes the point that contributions spill over from one of my arbitrary categories to another. The more intimate scale of work may be thought of as ecological, the broader area study as geographical and it is certainly founded on analytical work. Indeed I shall have to cross-refer again to this work in the next section since the problem-oriented work in U.S.S.R. is carried through to what one might call applied regional synthesis in medical geography.

So far the examples have concerned infections, and arthropod-borne at that. This is a distorted image of much of the literature cited, arising from the more immediate impact of ecological concepts to arthropod vectors of a human disease like malaria, or an animal infection in which man becomes involved, like scrub typhus. This can now be corrected by reference to studies of diseases not known to involve infections or arthropod vectors.

On quite a local and intimate scale was the now classic study by Allen-Price of areal differences of cancer incidence in relation to different sources of water supply. Of course yet another boundary between disciplines is involved, and to follow up this work one has to range into geological literature.[7]

It seems clear that maps were used as tools of analytical value and not as mere illustrations (Fig. 2.2):

Having investigated the cancer incidence in a general statistical way, which did not yield any really positive result, except those set out in the tables, the situation was plotted on a map. The 5547 deaths were depicted on a 6 in. to the mile Ordnance Survey map. Deaths from cancer were marked by a cross, and those from other diseases by a circle. The extraordinary distribution of the disease at once became apparent. For example, in one hamlet the ratio of crosses to circles is 1 in 12, whilst in an adjoining hamlet of comparable community the ratio is 1 in 3 . . .

Fig. 2.2, showing the distribution of deaths from cancer and from other causes in Horrabridge, is diagrammatic and has had to be reduced from the 24 in. to the mile Ordnance Survey map [probably in fact the 6 in. to the mile map]. On this larger scale the individual houses and their water-supply appeared in distinctive colour, with the cancer and other deaths. It is not possible to convert this colour presentation to black-and-white symbols and still maintain complete accuracy. For example, the line of demarcation between the sectors

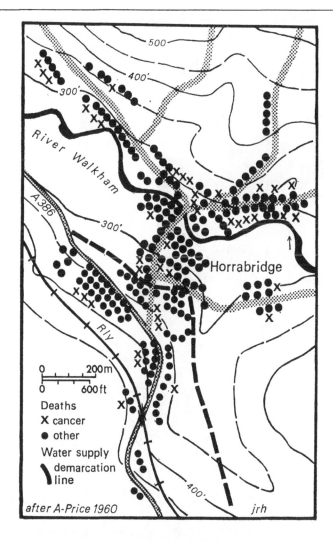

FIGURE 2.2 Cancer deaths and deaths from other causes mapped in relation to three sources of water supply (i) north of the River Walkham, (ii) between the Walkham and the pecked line and (iii) south and west of the pecked line. (From: Allen-Price, E. D. (1960) *Lancet*, i, 1235, Fig. 2, p. 1237.)

is too sharp and makes no allowance for infiltration of the Plymouth water-supply to the south-west and the Watery Ford supply to the north-east. But Fig. 2.2 does show the remarkable number of crosses representing cancer which lie to the north of the River Walkham.

The distribution of cancer in this compact community of Horrabridge which in the radius of half a mile has three distinct mortality-rates, cannot be explained by any current theory. Here there is a homogeneous group, following

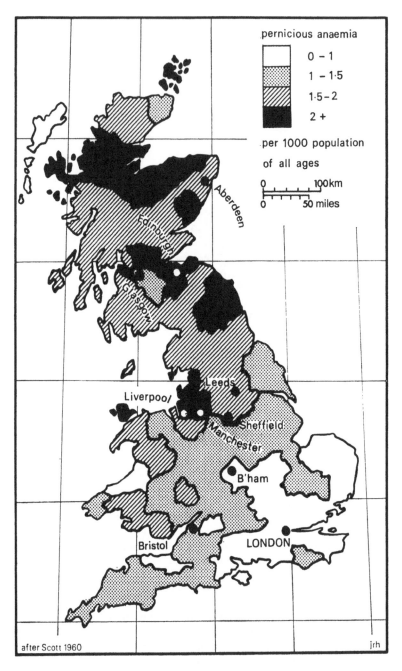

FIGURE 2.3 Pernicious anaemia prevalence mapped from data from general practice.
(From: Scott, E. (1960) *Journal of the College of General Practitioners*, 3, 80.)

the same occupations, eating the same food, and in an identical environment, and merely separated one from another by the natural boundaries of the River Walkham, which is crossed by a bridge. Here, for generations, the people have inter-married freely, and their social activities have been combined, yet each artificial section of the community has a widely different cancer mortality. As far as can be assessed the only difference that could account for this is their water-supply.

On a country-wide scale, the opportunity offered by the research register of the College of General Practitioners was taken up in an interesting study of pernicious anaemia in Great Britain (Fig. 2.3).[8] The discussion reads:

The regularity of the gradient in prevalence of pernicious anaemia from one part of the country to another was unexpected. The problem now is to try and determine its causes. Pernicious anaemia appears to be more common in people with blood group A than in those of other groups. Maps now being made by Dr. Mourant of the Nuffield Blood Group Centre to show the distribution of the A, B and O blood groups in the United Kingdom, however, indicate that group A is least common in the north and west, where pernicious anaemia is most common and vice versa. Inequalities in the distribution of blood group A, therefore, cannot be the cause of the variations in prevalence of pernicious anaemia revealed by our survey.

Professor L. J. Witts has suggested that the variations in prevalence of pernicious anaemia now may match variations in the rate of unemployment in the 1930's. This hypothesis is being tested more closely.

Another possible correlation is with the geological structure of each area. It is to be noted that in general the rates are low where sedimentary rocks are to be found and highest where the rocks are predominantly volcanic.

Clearly a great deal more work must be done before any useful answer can be given. Meanwhile those who have contributed to the survey can justifiably feel that their efforts have served to break new ground.

The survey has also shown that ways can be found to collect from general practitioners valuable information about diseases or other conditions, which are well-defined. Experience in this enquiry suggests that, in any future survey, time should be taken to choose the method of collecting information best suited to the purpose. It will probably be wise for one person to analyze all the material, even though several subsidiary collecting points are set up. For example, several faculties of the College made particular efforts to secure a good response from their members; in other counties the best results came with the help of executive councils. All the reports, however, were analyzed centrally by the same person.

This study is interesting because it led to geographers trying to improve the mapping and cartographic correlation (Fig. 2.4). This collaboration was interrupted by the death of Dr. Scott, and the map is hitherto unpublished;

it seemed to me unwise to put it out as a contribution to knowledge about the disease without further medical collaboration, especially since there are difficulties about the estimation of sampling error in volunteers for the Research Register of a College of which membership is optional. My own opinion is that we need not anticipate bias in the disease patterns in the practices of the record keepers, but there may be areal differences in the

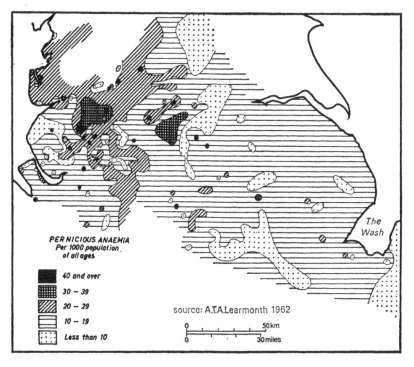

FIGURE 2.4 Pernicious anaemia prevalence in a belt across the middle of England, mapped from the same data as Fig. 2.3, using values from individual practices as 'spot-heights' and interpolating isopleths. (A. T. A. Learmonth/E. Scott.)

total representation of doctors in the sample which remained untested when the study was abandoned.

Studies of eye diseases in Australia, a country of continental dimensions, culminated in a world survey specifically geographical in intent, like many of Ida Mann's papers too numerous to cite here.[9] It is difficult to give the flavour of her writing but the following summary of an oral paper[10] may whet the appetite:

Some geographical pointers in the study of the distribution of eye disease.

Human diseases can be divided into groups, congenital, traumatic, inflammatory, neoplastic, degenerative and functional. In many instances a definite connection with some aspect of the geography of an area can be demonstrated. Geographical features which influence eye disease include isolation on islands or in inaccessible regions (leading to inbreeding and the emergence of congenital defects), excessive amounts of ultra violet light (producing through trauma snow blindness, pterygium and basal celled carcinoma of the lids), conditions of temperature and humidity favourable to the breeding of insect and other vectors (of diseases such as tularaemia, and malaria), altitude (hindering the growth of bacteria, insects, e.g. Anopheles), local peculiarities of flora and fauna (accounting for injuries and allergies or direct attack, as by spitting cobras), and the suitability of the soil for cultivation of different foods which may influence general nutrition and induce disease by deficiency of trace elements (for example, iodine lack and cretinism).

On the world scale a contribution from many aspects of a major public health problem is a W.H.O. symposium on goitre (W.H.O. 1960). The book as a whole is dominantly by medical men, but interestingly enough the chapter on 'Prevalence and geographical distribution' is by chemists of the Chilean Iodine Educational Bureau in London[11] though not that on technique of endemic goitre surveys.[12]

The increase in goitre in Tasmania in the early 1950's is a particularly fascinating story, and one still producing surprise pieces of new information in 1970. The early studies were largely medical and statistical, but might so well have been geographical that they are worth mentioning.[13] The basic problem of the origin of goitre is still unsolved, and it is possible that mapping and cartographic correlation on a variety of scales might yield results given collaboration with a geochemist.

Some of the studies flowing from May's *World Atlas of Diseases* have already been mentioned. His edited volume of *Studies in Disease Ecology*[14] include several that are appropriate to this section, for they progress from the local ecological web to the broad areal distribution pattern. They include May's own paper on malaria and an important one of Audy's series of papers on scrub typhus discussed above.[15] May's series of volumes on the ecology of malnutrition in the underdeveloped world represents an individual contribution of staggering magnitude to one of the major problems of our times.

Similarly the *Welt Seuchen-Atlas* stimulated a flow of papers even apart from the scholarly memoirs on the areal distribution patterns of disease on a world, continental or sub-continental scale in the Atlas itself.[16] From Rodenwaldt, for example, came contrasting papers, like that analysing the distribution of Chagas' Disease across Latin America, and that on the 1576 plague epidemic in Venice.[17] Similarly varied is the flow of papers from Jusatz and his colleagues at the Geomedical Research Unit of the Heidelberg Academy of Sciences, ranging across all three of our sections. Examples are Jusatz on the implication of successive seasonal peaks in different climates for bioclimatic classification,[18] Diesfeld on amoebiasis in tropical highlands[19] or Kauker on the world distribution of anthrax.[20] This could certainly be matched from the U.S.S.R. Perhaps the best point of entry to the Soviet literature is the medical geography section of the journal of abstracts (*Referatovayi Zhurnal*); study of a cross section of abstracts in translation suggests a strong emphasis on detailed working out of the ecological ideas mentioned earlier, usually in relation to application in relation to the regional ecology of an area and therefore relevant to the following section on regional syntheses. There is a strong concern with methodology, as in the volume from the research institute for Siberia at Irkutsk on principles and methods of medical cartography.[21]

On a broader scale of time and place, medical geography has its cultured sister, medical history. The vast literature there has often some relevance. Longmate's *King Cholera*,[22] following something of the tradition of Zinsser's *Rats, Lice and History*,[23] complements Pollitzer's W.H.O. monograph of 1959. Stimulus is surely to be gained; consider this passage from Henschen's fascinating *History of Disease*:[24]

> The appearance and character of diseases are subject to historical development and varying geographical and demographical conditions of population. Some diseases seem to disappear; other new ones to appear. Infectious diseases . . . have been driven back by the advance of medicine. Instead two other groups of diseases, cardio-vascular diseases and tumours, have taken the first place. . . . The overall picture of diseases within one country or community, which one can call the 'disease-panorama', varies . . . from time to time, from country to country, and from town to town.

Lastly and different in kind, Maegraith's crusade on exotic diseases of the last five years or so is less directly geographical. His interest is to get general practitioners to ask a patient with unfamiliar symptoms if he has travelled in areas with a different disease-complex. Air travel adds urgency to the plea.[25] Simple cartography is used, but there is geographical potential, perhaps even in relation to probability mapping of possible flows of infections. The theme is clearly important and it adds powerful impetus to the type of work outlined in the last section of this volume, which geographers might otherwise neglect in the current flight from regional geography towards 'the new geography'. Maegraith's case does not rest solely on the interests of the developed countries:

> The redistribution of exotic diseases and their introduction into otherwise free or relatively free areas is equally important in many of the developing countries themselves and requires the same kind of understanding of the geographical distribution of disease and awareness on the part of the practising doctors and health authorities. There are few large-scale socio-economic developments which are not involved in this way, as the result of movement and re-grouping

of local populations, migration of labour from other areas of the country or abroad, and the intermingling of local and immigrant populations with the associated medical risk of interchange of infections and exposure of non-immunes. This may be seen in ventures such as the Volta dam in Ghana which attracts labour from all over the country, and the extension of road systems for opening up previously isolated areas, as in the north-east of Thailand. The ultimate success of these and similar projects in no small degree depends on the redistribution of disease and the control that can be exercised upon it, since in the long run the health of the economy of a developing country will be determined by the health of its community.

REGIONAL SYNTHESES

Maegraith's special plea apart, May's characterisation of interrelationships between men and infections in Viet Nam has already made some of the case for all-round understanding of the disease complex in an area.

Jusatz and his colleagues from Heidelberg have recently revived and given fresh vigour to regional memoirs on medical geography: a substantial and scholarly Geomedical Monograph Series of Regional Studies in Geographical Medicine began with Kanter on Libya,[26] Fischer on Afghanistan[27] and Schaller on Ethiopia.[28] The books are in both German and English. The project engages my sympathy and support since my own first substantial paper on medical geography was on comparable ground in relation to the former British India for the inter-war period, the last reasonable span of years without interruptions in the flow of data before the end of the colonial period.[29]

It is extremely difficult to give anything of the content or even the flavour of a substantial monograph in a short extract, but something of the approach may be derived from the following, part of Kanter's last chapter entitled 'Region and disease: geomedical conclusions':

We know the shifting epidemics (Jusatz) as opposed to nestling epidemics, which are related to the region independently of human agency. In an endemically infected area with sporadically occurring cases, isolated foci of the latter come into existence according to climate and soil, the development of the pathogenic agent and of the transmitting vector, from which the epidemic outbreak concerned may be exported and spread by man or animal. Geomedical conditions in Libya enable shifting epidemics to cover a large area. A few examples may help to illustrate this.

Cases of typhus caused by rickettsiae for instance, chiefly occur during the cold season, when warmer clothes are worn and kept on the body during the night as well. This humid microclimate of the human body, averaging between 31·5 and 33°C, suits the transmitting lice remarkably well. They then deposit their eggs (nits) which remain on the body hair and clothing. Five to six days later the young lice hatch. The warm, dry season is less favourable to the lice as the humans are then wearing light clothes and the lice are exposed to high temperature and insolation. In areas with an insufficiently hygienic way of life, to be met amongst the nomads in Libya, in Cyrenaica and especially in the Fezzan, there are small epidemics almost every winter with fatal results among

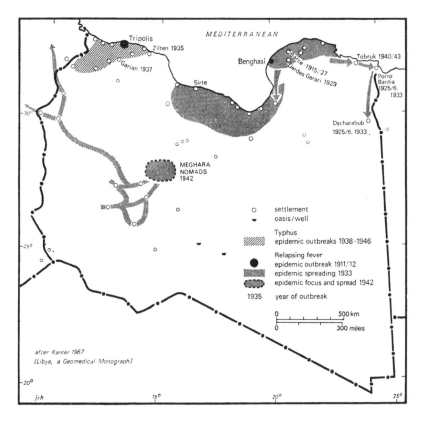

FIGURE 2.5 Libya: epidemic outbreaks of typhus and relapsing fever, 1911–46. (From: atlas section, Kanter, H. (1967) *Libya: a Geomedical Monograph*, trans. Hellen and Hellen, Geomedical Monograph Series: Regional Studies in Geographical Medicine, New York: Springer-Verlag.)

the children particularly. Most children, however, become immunized for life. The epidemic of the period 1938–46, which appears to have originated in Garian (Fig. 2.5) will be taken as an example to illustrate the participation of the nomads as transmitters of the disease.

Lice as well as certain kinds of ticks also spread borreliae which cause relapsing fever. Sporadically occurring cases in northern Africa leave the question open on whether the fever is to be considered as caused by lice or by ticks. Louse-borne relapsing fever is the European form and also occurs occasionally in North Africa and in tropical highlands. Ticks (*Ornithodorus*) as domestic pests live in clefts and cracks of houses and in sandy soil. *Borrelia* may remain virulent in the ticks for years and even transfer from infected ticks to the eggs, which are laid in the soil in dry places. Transfer occurs between rodents and man via the ticks and vice versa, as well as from man to man. The indigenous population acquire a high degree of immunity against the disease.

In the transitional areas of the countries of the arid climatic zones and in the oases of the desert, *Ornithodorus* lives far from man in the holes of rodents and foxes, infecting them. Consequently these areas must be regarded as permanent foci for this spirochaetosis which does not therefore seem to be exterminable. Men become infected when sleeping near the holes of rodents, or on a tent-place, in a cave or well infested by ticks. The nomads in turn bring the ticks to relatives and friends or they are carried off together in the camel bales as happened in 1942 to 1944, when the disease advanced to Tunis, Algeria and even further with the caravans then operating. In 1933, too, nomads brought the disease from Cyrenaica to the Sirte, and there was an epidemic in the city of Benghazi; in 1911/12, an epidemic swept through the city of Tripoli.

Finally plague should be mentioned as a shifting epidemic, although really it ought to be considered a nestling disease. In the course of centuries, plague was imported again and again into Libya either from overseas or by land routes. Epidemics in towns can generally be traced back to infected rats, which, originating in the ports, infected the rats of the towns, and thus the people too, with their fleas. From here the urban form spreads, carried by the semi-nomads, to the rodent fauna of the steppe and semi-desert, thus leading on to the infection of large areas where small epidemics of rural form occur in native settlements. Again, rodents are probably responsible for spreading the disease from the large Egyptian focus to Cyrenaica.

Fleas as well as their larvae are very sensitive to extreme aridity; they require a certain degree of humidity. The further towards the desert they find them-selves transported, the more they are affected by aridity and the more does their mortality rate increase. The Sirte areas seem to present a border zone for them just as the total desert prevents the fleas from advancing further to the south and thus taking the disease into the Fezzan. As motor vehicles increase in numbers this desert boundary can be crossed more easily, and fleas may well spread on the rodents of the oasis should plague enter the area once more as a shifting epidemic.

In Australia a medical man turned statistician has followed up Harvey Sutton's pioneer and in a sense premature paper to geographers in 1933:[30] Lancaster's 'Vital statistics as human ecology' of 1963[31] is a national rather than a regional synthesis, published towards the end of a long series of mortality surveys. Geographers may find the following extract challenging, even though areal distribution patterns are not at the heart of the discussion:

Many infective diseases affect young children and infants, a feature which some texts state is characteristic of these diseases. Let us consider in detail the epidemiology of measles.

Measles is a virus disease with an incubation period of about 10 days if measured up to the first signs of catarrh or 14 days up to the appearance of the rash. Let us agree to say that the length of time between cases in any chain of infection is a fortnight. It is very contagious under favourable conditions, sometimes attacking 100 per cent of the population. Measles epidemics of this explosive type have been reported from the Pacific Islands in the last century and recently from Greenland, the Faeroes and other isolated areas. Australia has also had some disastrous epidemics: for example, in Victoria 1797 deaths from measles occurred in 1874 and 1875. This epidemic also came as a great surprise to the commentators on Australian mortality for they caused the crude death rate to have a far greater variance than in England. It would be easy to postulate a special strain of the disease but let us consider the problem from an ecological view.

To maintain such an infection as measles in a community, at least one infection every fortnight is needed to keep the epidemic going. If conditions are such that only one or two occur each fortnight, there is a definite probability that some chance factor will intervene and the chain of infections will be broken, so that we shall need, say, an average of ten new cases a fortnight or 260 a year. But this corresponds to a minimum population of the order of 10,000. Now the epidemic cannot direct itself and from time to time there will be chance spread to hundreds of cases, thus needing a larger population. Spatial considerations point to further difficulties for the maintenance of the disease because each sub-population – for example the inhabitants along the banks of a coastal river of New South Wales – were not only effectively isolated from those of other regions but also they were broken up into smaller subaggregates, such as townships or farms. Measles, therefore, tended to die out in Australia. For example, there were 48 deaths in Western Australia in 1861 and no further deaths from measles until 1880. Similar series could be quoted from the other States. Such isolation and consequent dying out of the epidemic enables a large number of susceptibles to accumulate in the population and, when the disease is reintroduced, a large epidemic occurs attacking persons of all ages. So that in the Australian type of experience, large epidemics of measles occur after irregular intervals of many years. Measles epidemiology can be followed in Australia because it is a killing disease and deaths have been recorded (since it is a killing disease absence of measles deaths can be interpreted as absence of measles infection) and because it was compulsorily notifiable in the past. However, the same is not true of rubella, a disease with negligible mortality and once thought to be of no importance other than its possible confusion in diagnosis with measles. However, this view was changed by an observation of N. M. Gregg in 1943. Gregg was able to report 75 cases of congenital cataract in children whose mothers had had rubella in the early months of pregnancy. He noted also other congenital defects. C. Swan and his co-workers in Adelaide, however, first reported congenital deafness. After Gregg's and Swan's observations, theorists were not lacking in Australia or abroad who postulated that the epidemic was due to a new strain of rubella as the production of congenital defects by the virus was believed to be a new phenomenon. Let us look, however, to see whether simpler explanations are available.

In his Report on the Australian Census of 1911, the Commonwealth Statistician had noted the extraordinary concentration of deaf-mutes in the age groups 10 to 14 years. Thus in New South Wales there were 59 deaf-mutes at ages 5 to 9 years, 111 at ages 10 to 14 years and 64 at ages 16 to 19 years. A similar finding was reported at the Census of 1921 and again at the Census of 1933. The lack of regularity in the times of the Census allowed a guess that those born in 1899 or 1900 were responsible for this finding. All admissions to the Institution of the Deaf, Dumb and Blind at Darlington were therefore classified by date of birth. Of the admissions 15 had been born in 1898, 70 in 1899 and 16 in 1900, strongly suggesting some epidemic incident. Closer analysis by month in 1899 showed that there had been born 1 in January, 5 in February and 2 in March; and throughout the rest of the year 8, 8, 8, 14, 12, 6, 1, 2 and 2 which, plotted as a graph, give a definite epidemic wave. A search of the medical journals showed that there had been an epidemic of rubella in October 1898, the preceding year. Moreover, the birth numbers showed that there had been an excessive number of births of the deaf in 1916, which also had not been previously recognised and in 1924 and 1925, from which one case had been previously reported; all except one of Gregg's cases had been born in 1938, 1940 and 1941. Similar results were obtained from institutional data in the other Australian States and from New Zealand. It was clear that other isolated areas might have a similar experience. Data from Hawaii were not available but the Icelandic figures were of interest. There are usually one or two children born each year in Iceland who are later admitted to the institution for the deaf. However, there were 2 in 1939, 2 in 1940 and 10 in 1941, whereas in the three succeeding years, there were 2, 1 and 1 respectively. In 1941 the births occurred, 1 in January, 1 in May, 2 in June, 1 in July, 1 in August and 4 in September, once again suggesting an epidemic wave. Iceland is unique in having notifications going back many years (to 1880 in fact, although the earliest records are incomplete). Since my survey J. Sigurjonsson (1961 and 1962) has reported epidemics of rubella caused deafness for 1899, 1916, 1924, 1940 and also cases of congenital cataract. From his work, it is evident that every large epidemic of rubella in Iceland has led to cases of deafness and every epidemic incident of deafness has been preceded by an epidemic of rubella. The experience of England, United States of America, Sweden and Italy were quite different. Only after 1936 was there any tendency for deafness to have an epidemic incidence in England and Sweden: minor epidemics were reported in the United States in 1898. The implications of this work are clear. The property of producing congenital defects is probably a fundamental property of the rubella virus and not necessarily one of a particular strain. Congenital defects occur as a result of infection with rubella only in demographic conditions where the females can pass through childhood without acquiring the disease. Under such conditions, epidemics occur in which females of childbearing age are infected, some at the appropriate time of their pregnancy when the organs of the foetus are being laid down and a proportion of such foetuses are born with congenital defects. The epidemics of congenital defects are thus explained without the need for any additional hypothesis.

Measles has a higher case fatality rate (i.e. deaths/cases) in infants than in older children. Any demographic or social event which delays the age of infection is thus effective in bringing down the mortality rate in the population. The age at infection with measles in the past had a marked social gradient. Thus in Glasgow it was found that children of parents living in blocks of flats (lands) acquired the disease usually in the first year, those in industrial suburbs somewhat later, those in residential suburbs when they went to school and those who went to residential schools acquired the disease there often in their teens. There are no doubt many such ecological factors acting in a human population – some of which cannot readily be determined.

Differences within Australia are at the heart of the problem in Billington's series of papers on changes in the incidence of gastric ulcers.[32] A short extract from his 1965 paper the geographical reader may find intriguing:

> I have previously drawn attention to a change in the pattern of gastric ulcer which has occurred in Australia since 1940. The change involves an increase in the national mortality, coinciding with an increased number of hospital cases of gastric ulcer in women of child-bearing and premenopausal age. This phenomenon appears to have no parallel in other parts of the world and the available evidence suggests that altered environmental conditions have operated to produce ulcers rather than to prevent healing or promote chronicity. Figures from hospital in-patients in Sydney indicate that this change in pattern commenced in 1943.

> This paper is a further instalment in the unfinished narrative of the Australian gastric ulcer change. The story to date has been summarised elsewhere but briefly it relates to a notable rise in the incidence of chronic gastric ulcer in women in New South Wales and Queensland (but not in the other States of Australia). This appeared to start early in the decade 1940–49, first affecting women of child-bearing and premenopausal years, but it later involved women of older ages in both city and country areas, at least in New South Wales.

> The changing situation of gastric ulcer in Australia has produced a remarkable disparity in the male-to-female ratio for the disease between the two largest cities – 1·8 in Melbourne and 0·6 in Sydney.

> It has always been assumed that there is little variation between the States of Australia with regard to the genetic constitution, social structure, and habits of the population; it is curious that such a pronounced difference between the States in the pattern of gastric ulcer in women could have occurred. The reason is unknown, but it was suggested that study of border areas between affected and unaffected states might provide information leading to a solution.

New South Wales has borders with two States in which the gastric ulcer change has not occurred – Victoria and South Australia. In New South Wales on the border with Victoria lies the rural city of Albury (population 23,520) which is known to be linked socially and economically much more closely with Melbourne than with Sydney.

I carried out a retrospective study of gastric ulcer in Albury and the results of this study led to a similar investigation in the New South Wales rural city of Wagga Wagga (population 24,840), eighty miles north of Albury.

Wagga and Albury are the two largest and most important rural cities of the south-western Slopes district of New South Wales, set in the rich wheat and sheep country of the eastern Riverina, so named because of its relationship to the western watershed of the Dividing Range, comprising the Lachlan, Murrumbidgee and Murray rivers. While Wagga is situated on the Murrumbidgee river, Albury is on the New South Wales side of the Murray river which, as it flows westwards, forms the boundary between New South Wales and Victoria.

Between Albury and Wagga are six small towns with a total population including their surrounding districts, of 5281. The economics of both Albury and Wagga depend on the same rural industries. Consumer goods, except those primary products farmed locally, come to Albury mainly from Melbourne and to Wagga from Sydney.

There exists a clear difference in the male-to-female ratio between Albury (1·9) and Wagga (0·6) with respect to gastric ulcer. It is notable that the pattern of gastric ulcer in Albury reproduces that found in Melbourne, while the pattern in Wagga is similar to that in Sydney and the rest of New South Wales. It seems reasonable to conclude that the geographic demarcation between the area affected by the Australian gastric ulcer change and the unknown factor responsible for it, and the unaffected area, lies in New South Wales between Albury and Wagga. It was not possible at Wagga to scrutinise radiological records for the period 1946–60 to see whether the rising incidence of gastric ulcer in women was contemporaneous with the events in Sydney, nor indeed to determine at what period in the last twenty years the changing pattern of gastric ulcer was first seen in Wagga.

As Albury in New South Wales retains, with Melbourne, the sex and age distribution of patients with chronic gastric ulcer existing in Sydney and Brisbane before the Australian gastric ulcer change began, the inference is that the factors responsible for the change are not related to differences in legislation and regulations between the States as had been previously suggested. Inquiry into commerce and the distribution of consumer goods seems more likely to reveal differences which could lead to the isolation of the factor responsible for the production of so many gastric ulcers in the women of the rest of New South Wales and Queensland.

Conditions in the major Australian cities, Sydney and Melbourne are difficult to investigate, but it should be possible to find differences between Albury and Wagga as they are dependent on outside sources for most consumer goods.

Theoretically the only possible approach seems to be to scrutinise personally every detail of the entire life pattern of women, searching for unknown differences in the behaviour or in the types and sources of the commodities used, using non-ulcer controls from Albury and comparing them with newly diagnosed index cases from women with ulcers in Wagga Wagga – a formidable task.

On a supra-continental scale Banks, in 1958, addressed a research agenda to geographers in terms contrasted for the developed and for the under-developed world.[33] After an account of health in Britain while it was as yet underdeveloped, to use the modern term, he went on:

The first half of the twentieth century has witnessed more dramatic advances in the medical and social fields than at any other time in our history. We, in this country, have experienced three major wars within fifty years, each of which has resulted, directly or indirectly, in medical improvements. The Boer War, with its deaths from disease, and especially from enteric fever, together with the startling disclosure that many of the volunteers from the industrial cities were physically unfit to serve, acted as the spur. Within a few years of the ending of that war there began the introduction of medical and social legislation, including the School Medical Service, and culminating in the so-called Welfare State after the Second World War. Comparable developments have occurred in a number of other countries, where the vital statistics show similar improvements to our own and, indeed, are sometimes better than those to be found in Britain.

Perhaps the most remarkable change of all is illustrated by the fact that within my own lifetime, little over fifty years, the expectation of life from birth in England has increased by twenty-five years for men and thirty years for women. The biological balance in this and similar countries is now one of relatively few but healthy children, healthy adults, and an increasing number of old people.

The contrast between such favoured societies and other parts of the world is sharp indeed, for the expectation of life from birth remains at about thirty on the average in many parts of the world where the ancient disease pattern has undergone little change.

It would simplify matters considerably if we could make a sharp distinction between urban and rural communities, or between the industrialized and non-industrialized parts of the world. Unfortunately it is not as simple as that. So many other actors, including nutrition, play their part. Worthington drew attention to something of this kind when he said 'A visitor . . . cannot help but be impressed by the influence on native health and physique of good regular food and medical attention during a contract for work on the Witwatersrand Mines of South Africa. The contrast between the incoming and outgoing labour force must be seen to be believed' (Worthington, 1958).*

* Worthington, E. B. (1958) *Science in the Development of Africa*. Commissioner for Technical Cooperation in Africa South of the Sudan, and Scientific Council for Africa South of the Sudan, London.

I would like to suggest that more attention might be paid to the 'fringe' areas where different cultures meet, and also to those islands which form melting pots for the fusion of different racial groups. Field studies in such areas, made after careful preliminary preparation, could be most revealing.

More recently Jusatz has also written on this scale and on somewhat similar lines, both in German and in English;[34] there is a strand also of thinking parallel with Maegraith's, as we see from the conclusion to the paper in English:

In spite of good international co-operation in the fight to control epidemics through W.H.O., we in Europe, in the countries of the temperate climatic zone, still need to feel threatened for the following three reasons:

1. The speed with which ubiquitous epidemics can spread has increased:

   (a) because the greater flying speeds have shortened the times of inter-continental flights and

   (b) because of the increase of air traffic between the continents.

2. We have not yet succeeded completely in eradicating the agents of even a single epidemic disease. The breeding grounds of the great epidemics of old have not yet been rendered completely safe.

3. The causative agents of an epidemic can return from the tropics in a mutated form to the temperate zone, as is shown by the example of the resistant gonococci.

   Hence it requires the vigilance of the physicians of all countries, especially those engaged in the public health service, to watch out for the following: –

   (a) To keep the smallest incidence of epidemic diseases under constant observation.

   (b) To make full use of the radio-telegraphic warning system of the W.H.O. because the danger still exists that epidemics might break out and be introduced somewhere else.

   (c) To keep up the proven methods of individual and mass prophylaxis especially for air travellers, by vaccinating against smallpox, yellow fever and cholera, in accordance with the international health regulations as laid down by the W.H.O.

   The epidemiological situation in the world is still far from good, and a great deal of care and attention will be required to make any improvements, especially in the countries of the tropics.

But the paper in German contains a much stronger element using the closest integration of health planning with general developmental planning in underdeveloped countries; and the English summary concludes thus:

The experience that has been gained to date through the training of a special professional group, by means of a shortened medical course without the usual academic qualifications, and the setting up of so-called health centres in a number of African countries, appears to indicate that this is the right method, and that this method can lead to an improvement in the contagious disease situation, which must still be considered serious in the tropical countries.

CONCLUSION

I think that our editor was justified at the present stage in collecting a volume of papers in medical geography by professional geographers. As a whole they are likely to stimulate interest among students of geography, but also among medical and biological workers who may have missed the growing stream of work by geographers. This paper is the least likely to excite medical men, but may interest geographers approaching the field afresh. All the medical contributions cited have certainly aroused my own interest, spread over half a working life; they may appear bitty and ill-disciplined. Can a short conclusion do anything to remedy this?

I think I can only plead that the reader look to the convergences between sciences which have been so marked and so cheering in the last decade or two. The convergence between the biologist's ecology and medical geography is not the only such convergence in modern geography. Stoddart has pleaded for the analysis of ecological webs as a main unifying factor in modern geography, at a suitable level of sophistication in handling evidence.[35] Young geographers who have grown up with the new geography and with statistical and mathematical approaches which a much smaller minority of professional geographers possessed a generation ago are much more at home with the methods of analysis of an areal distribution pattern that would commend themselves to workers in medical statistics or biomathematics; men such as Doll,[36] Bartlett,[37] Bailey[38] and Rapoport,[39] to name but a few. Haggett,[40] while referring to medical geography as 'a confusing sub-variety', has in fact gone on to use several examples from medical geography in demonstrating methods of locational analysis in human geography – and he has agreed to contribute to this book. Few, if any, of the problems or approaches suggested in this paper, however tangential at first sight, cannot be foreseen as becoming confluent with the mainstream of spatial analysis of areal distribution patterns, of differing density, of flow or diffusion, of differing trends.

REFERENCES

1 MAY, J. M. (1950) *Geogrl. Rev.*, **40**, 10.
2 MAY, J. M. (1959) *Ecology of Human Disease*, New York.
3 AUDY, J. R. (1949) *A Summary Topographical Account of Scrub Typhus, 1908–1946*. Bulletin of the Institute for Medical Research, Kuala Lumpur, 1 (new series).

4 AUDY, J. R. (1968) *Red Mites and Typhus*. London.

5 AUDY, J. R. (1958a) *Br. J. clin. Pract.*, **12**, 102.

 AUDY, J. R. (1958b) *Trans. Roy. Soc. trop. Med. Hyg.*, **52**, 308.

6 AUDY, J. R. (1965) 'Types of human influence on natural foci of disease'. In B. Rosicky and K. Heyberger (eds). *Theoretical Questions of Natural Foci of Diseases: Proceedings of a Symposium*. Czechoslovak Academy of Sciences, Prague.

 PAVLOVSKIY, E. N. *et al.* (1955) *Natural Nidi of Human Diseases and Regional Epidemiology*. Medgiz, Leningrad (in Russian). Translation, as *Natural Foci of Human Infections*. Israel Program for Scientific Translations, Jerusalem, 1963.

 FORMOSOV, A. N. (Ed. 1965) *Fauna and Ecology of the Rodents*. Moscow Society of Naturalists, Proceedings on the Study of Flora and Fauna of the U.S.S.R. (new series). The Section of Zoology No. 40. Materials on the Rodents, No. 7. Publishing House of Moscow University. Academy of Sciences of the U.S.S.R., Institute of Geography of Siberia and the Far East, Siberian Branch, 1964. *Medical Geography: Outcome and Outlook*, p. 98. Irkutsk.

7 ALLEN-PRICE, E. D. (1960) *Lancet*, i, 1235.

 WARREN, H. V. (1967) *Ann. New York Acad. Sci.* **136**, 657.

8 SCOTT, E. (1960) *J. Coll. gen. Practnrs.*, **3**, 80.

9 MANN, I. (1966) *Culture, Race, Climate and Eye Disease: an Introduction to the Study of Geographical Ophthalmology*. C. C. Thomas, Springfield, 111.

 MANN, I. (1963) 'Trachoma in Australia'. *Orient. Archs. ophthal.*, **5**.

 MANN, I. (1959) 'Research into the regional distribution of eye diseases'. *Am. J. Ophthal.*, **47**, 134.

10 MANN, I. (1965) 'Some geographical pointers in the study of distribution of eye disease'. Summary of Paper, ANZAAS, Hobart Congress.

11 KELLY, F. C. and SNEDDON, W. W. (1960) 'Prevalence and geographical distribution of Endemic Goitre'. In W.H.O. Monograph Series, Geneva, No 44, p. 27.

12 PEREZ, C., SCRIMSHAW, N. S. and MUNOZ, J. A. (1960) 'Technique of endemic goitre surveys'. In W.H.O. Monograph Series, No. 44, p. 369.

13 CLEMENTS, F. W. and WISHART, J. W. (1956) *Metabolism*, **5**, 623.

14 MAY, J. M. (ed.) *Studies in Disease Ecology*. New York, 1961.

15 MAY, J. M. (1961) *The Ecology of Malaria*, pp. 161–229.

 AUDY, J. R. (1961) 'The ecology of Scrub Typhus'. In J. M. May (ed.) *Studies in Disease Ecology*. New York.

16 RODENWALDT, E. and JUSATZ, H. J. (1952, 1956, 1961) *Welt-Seuchen Atlas*. Hamburg.

17 RODENWALDT, E. (1959) *Zeitschrift Tropenmedizin und Parasatologie*, **10**, 1.

 RODENWALDT, E. (1952) *Sitzungberichte der Heidelberger Akademie der Wissenschaften*. Mathematisch-naturwissenschaftliche Klasse, 2 Abhandlung.

18 JUSATZ, H. J. (1963) 'Seasonal diseases and bioclimatalogical classifications'. Third International Biometeorological Congress, Pau (France).

19 DIESFIELD, H. J. (1965) *Zeitschrift fur Tropenmedizin und Parasatologie*, **16**, 401.

20 KAUKER, E. (1965) *Sitzungsberichte der Heidelberger Akademie der Wissenschaften.* Mathematisch-naturwissenschaftliche Klasse, 2 Abhandlung.
21 IGNATYEV, YE. I. (1966) 'Medical geography and practice'. In J. Ducolot (Ed.) *Mélanges de Géographie offerts à M. Omer Tulippe, Liège.* Gembloux.
VOROB'YEV, V. V. *et al.* (1968) *Principles and Methods of Medical Geography Mapping.* Irkutsk.
22 LONGMATE, N. (1966) *King Cholera: the Biography of a Disease.* London.
23 ZINSSER, H. (1935) *Rats, Lice and History.* London.
24 HENSCHEN, F. (1966) *The History of Disease,* p. 1. London.
25 MAEGRAITH, B. (1965) *Exotic Diseases in Practice.* London.
26 KANTER, H. (1967) *Libya: a Geomedical Monograph.* Springer-Verlag Geomedical Monograph Series: ed. H. J. Jusatz. Berlin–Heidelberg–New York.
27 FISCHER, L. (1968) *Afghanistan: a Geomedical Monograph.* Springer-Verlag Geomedical Monograph Series: ed. H. J. Jusatz. Berlin–Heidelberg–New York.
28 SCHALLER, K. F. (19) *Ethiopia: a Geomedical Monograph.* Springer-Verlag Geomedical Monograph Series: ed. H. J. Jusatz. Berlin–Heidelberg–New York.
29 LEARMONTH, A. T. A. (1958) *Indian geogr. J.,* **33,** 1.
30 SUTTON, H. (1935) *Aust. Geogrl.,* **2,** 3.
31 LANCASTER, H. O. (1963) *Aust. J. Sci.,* **25,** 445.
32 BILLINGTON, B. P. (1963) *Australasian Annals Med.,* **12,** 153.
BILLINGTON, B. P. (1965) *The Lancet,* ii, 378.
33 BANKS, L. (1959) *Geogrl. J.,* **125,** 201.
34 JUSATZ, H. J. (1965b) 'Die bedetung der seuchenlage für die entwicklung der tropenländer'. Sonderdrucke der Mitarbeiter Nr. 3, Sudasien Institut der Universität Heidelberg.
JUSATZ, H. (1965a) *Ethiop. med. Jr,* **2,** 272.
35 STODDART, D. R. (1965) 'Geography and the ecological approach: the ecosystem as a geographical principle and method'. *Geography,* **50,** 242.
36 DOLL, R. (1959) *Methods of Geographical Pathology.* Oxford.
37 BARTLETT, M. S. (1960) *Stochastic Population Models in Ecology and Epidemiology.* London.
38 BAILEY, N. T. J. (1967) *The Mathematical Approach to Biology and Medicine.* London
39 RAPOPORT, A. (1951) *Bull. maths. Biophysics,* **13,** 85.
40 HAGGETT, P. (1965) *Locational Analysis in Human Geography.* London.
HAGGETT, P. and CHORLEY, R. J. (1969) *Network Analysis in Geography.* London.

3 Mortality or Morbidity Mapping: Some examples from Yugoslav Macedonia

N. H. Copperthwaite

The functions of the medical geographer have been variously described, analysed, and commented upon,[1,2,3,4,] but there is one function which remains indisputably his own. This function is that of map-making, and much of the basic work in medical geography lies in the preparation of maps illustrating areal distributions of various diseases or groups of diseases. This function is in itself valuable, for in the hands of a skilled geographer, a spatial distribution can assume new relevance when carefully displayed upon a map, and hitherto undeveloped characteristics may be evinced from a seemingly random collection of figures. Of course the data which is used must relate to some form of areal distribution. Data about disease, in common with many other factors in the human environment, possesses this quality. It can therefore be mapped.

The disease map has already assumed a number of forms, and John Snow's single dot distribution map of cholera in London[5] is often cited as one of the earlier models to have proved its usefulness. Jacques May has taken the world view for his maps,[6] Howe in Britain examined the United Kingdom, and produced choropleth maps to illustrate the distribution of a number of diseases in that country.[7] Learmonth has used a variety of techniques, including composite maps in his work on India[8] and isopleth (or isomort)[9] maps of mortality in Australia.[10] Each of these, however, have used as their basis a mortality statistic, and there has been relatively little consideration of morbidity as a feature that could be mapped. Yugoslav statistics, which are not published, but to which the author has been allowed access, offer a starting point from which a consideration of both mortality and morbidity maps may be made.

The aim of the medical geographer in preparing his maps must be principally to illustrate the spatial distribution with which he is dealing as accurately and consistently as he can, bearing in mind that the map is only a tool to be used discriminatingly in the effort to overcome some of the questions posed by disease. It should therefore possess certain qualities which aid this function. It follows that some diseases may best be studied through the mortality map, whilst others demand a morbidity map, and we may make a basic division initially between those diseases where the result of contraction of the disease is death and those where the contraction of the disease does not necessarily, nor indeed often, cause death, and each may be differently dealt with.

As examples, two groups of diseases have been selected amongst those which occur in the Yugoslav Republic of Macedonia; they are influenza (I.S.C. 480–3)[11] and the combined group, chronic rheumatic heart disease (I.S.C. 410–16) and arteriosclerotic and degenerative heart disease (I.S.C. 420–2). Whilst the latter claims about 9,000 victims each year in Macedonia, only about 250 people die of influenza. However influenza is a very common complaint about which a great deal is known. The effects of it are important, if only in economic terms because of the large numbers of people involved. It is therefore illness rather than death which is of greatest significance, since even the mildest disease will in any case result in death of the weak, infirm or aged. The distribution pattern of influenza may indicate areas where the morbidity level is high in comparison to other local areas and there is therefore doubtless a case to be made for the medical geographical examination of those areas to try to ascertain the reasons for the high level, if they are not already known or apparent, and perhaps to attempt a 'cure'. A morbidity map of influenza would pin-point the problem, and perhaps suggest the reasons for it.

A mortality map of influenza would have a different function. If the distribution of those susceptible to influenza were uniform throughout the given area, then the mortality map would merely reflect the morbidity map at a lower numerical level, because the same proportion of those who contracted the disease in each local area would die. The assumption, however, is unlikely, and we may thus postulate that peaks of mortality indicate areas of special sensitivity to influenza. Thus when used in conjunction the two maps would be of special value. A flat or uniform morbidity distribution with peaked mortality distribution would accentuate the mortality pattern because persons in the peaked area, although no more likely to contract influenza than the rest, would be the more likely to die.

The second group of diseases are less well understood than influenza, and because they are generally confined to the older age groups (55 onwards) they are less important in economic terms but probably more so in social terms, especially since the age of 55 in Macedonia is the statutory retiring age, although, of course, the majority do continue to work. In Macedonia the two groups of diseases account for 12 per cent of all deaths each year, and so although they may be crippling diseases, their terminal nature is more immediately important. The mortality map would therefore seem to be the most appropriate initially. However in the search for cause and effect the use of the two types of map in conjunction may still be important in this example, and its value can be assessed in the following pages.

THE USE OF YUGOSLAV STATISTICAL MATERIAL

(1) *Morbidity*

The treatment of illness in Yugoslavia is centralised in clinics of various descriptions to which patients are expected to travel if they are fit enough to do so. The majority of clinics are located in the larger population centres and serve the population in the immediate area. The *Organizacione jedinice opšte praksa* has the majority of general medical service clinics dealing with day-to-day cases and it is at the *opšte praksa* clinic that most primary diagnoses are made. They are supplemented by specialist clinics which deal with workers, school children, pre-school children and maternity and gynaecological cases. Treatment is free, and although specialist clinics are not uniformly available, the general service is.

There are no formally devised boundaries to each clinic's sphere of influence, or catchment area, so that a patient may not necessarily attend the clinic which is geographically nearest to him, if it is more convenient or easier to reach another one, for example, because it is on a direct bus or rail route. It seems unlikely that there is a gravitational effect towards particular clinics by reason of their reputation or methods of treatment, as is the case in some of the less developed parts of the world, and in most cases patients are dealt with at the local clinic.

Half-yearly returns are made up at each clinic, showing the number of cases of each disease which have been treated during the six-month period and the number of new cases diagnosed. Thus the return will indicate the number of cases which have occurred within the catchment area of the clinic in that period. Only the primary diagnostic visit is recorded for these purposes, subsequent visits for further treatment being recorded elsewhere. The 'C' list of the International Classification of Diseases, Injuries and Causes of Death, 7th Revision[12] is used as a basis upon which the illness is classified, although there are certain changes made to this list in respect of categories of illness which are more relevant to the specialist clinics for pre-school children, and for mothers. The fact that the first visit to the clinic is the one which is recorded is most important because if subsequent visits were to be noted, this would give a false impression of the incidence of disease at that clinic.

No other information is available about the patients who are treated. In particular male patients are not separated from female, except in the obvious categories of disease, or where there are specialist clinics for women. Only very crude estimates can be made about comparative numbers in different age groups by using figures from the specialist clinics, but this is very unreliable since specialist clinics for, for example, children at school, or of pre-school age, are not uniformly available, and by far the majority of cases are dealt with at the general service clinics.

The material which is available, however, can be used in a number of ways. In its crude form it serves to indicate levels of demand for medical services in the various centres, it assists doctors and administrators in deciding upon the most rational utilisation of their resources, and it demonstrates the comparative health hazards in the community. More sophisticated use of the figures is, however, difficult, due largely to the fact that it is difficult to relate the number of cases of a disease diagnosed at a particular clinic to a particular population level. The 'rate', expressed as cases per thousand population, would be the simplest way of standardising the figures from each clinic and making them comparable with those from other clinics, but this presupposes that the number of cases may be related to a given population. Due to the difficulty of defining the catchment areas, the population of the particular clinic may be only subjectively assessed, with the possibility of large errors. A rate, however, is very suitable for mapping but difficult to apply in this case. The crude numbers themselves may be expressed on the map in a number of ways. For some diseases, especially those which may reach minor epidemic proportions, the overall view of disease unrelated to population may be all that is required.

Unless, for particular reasons, finite boundaries are necessarily required, the construction of an isopleth map, once data has been collected, may well be the most satisfactory method of illustrating that data. Areas of high and low morbidity will be evident and the gradient between the two can well be expressed. At the same time areal boundaries, which may be misleading, are not shown.

In order that high morbidity in real terms may be established it is obviously necessary to attempt some form of standardisation which does not rely on a base population of known characteristics. Also the decision must be made as to whether data is to be standardised against some international, national or, in the Yugoslav case, the Republican* criterion, although the choice will often be dictated by the material available. In the case of Macedonia, a local Republican norm was accepted, and the total number of primary diagnostic visits in the Republic was the base level. The total number of cases at each local clinic was then expressed as a percentage of all the cases in the Republic. Thus Clinic X, during the three-year period 1966–8, had treated an average of 2,000 cases in all per annum. If there were in all an average 200,000 cases in the Republic each year, Clinic X had a 1 per cent share.

The assumption upon which the standardisation procedure rests is that since the clinic has 1 per cent of all new cases in the Republic each year, it

* Six Republics make up the state of Yugoslavia.

will also have 1 per cent of all new cases in each disease. In practice this rarely occurs and the clinic may have only 0·5 per cent of the new patients with a particular disease. In that event, it would clearly have only half the number of cases which would be expected if the assumption were correct. Expressed in terms of base level 100, the local clinic had 50. If it had had 2 per cent of all the cases of a particular disease expressed in the same way, its standardised index would be 200.

This is not the only method by which the material could be standardised, but it has the advantage that although errors may be present, the comparative results in the Republic are unchanged, so long as all local calculations are made with reference to the same base level. It should be stressed that results thus obtained are comparable only at Republic level, and not with results obtained in the same way for other Republics in Yugoslavia or for other countries.

The method which has been described is the one which has been used to calculate the relative morbidity expressed in Fig. 3.4 and Fig. 3.6 of this paper. The total number of cases of each disease in each area has been arrived at by the addition of the cases treated at the general service and specialist clinics in the same town. All the communes in Macedonia have at least a general service clinic, and a few have two. Where there were two clinics in the same commune, they were treated as one unit for these purposes, so that a final figure for each commune could be obtained. There are obvious objections to this procedure, but for the sake of clarity and simplicity in this paper, and in the view of the fact that only two clinics were involved, the procedure was adopted. All the clinics in Skopje (see 'S' in Fig. 3.2), the Republic capital, were treated as one unit. The material was in this way made comparable to that for mortality.

(2) *Mortality*

Basic to the production of a satisfactory mortality map is the availability of satisfactory statistical material. In Yugoslavia all types of data are collected and collated at the Statistical Offices in the capital cities of the six Republics, and apart from National Statistics, which are gathered in Belgrade, all published data originates from these offices.

The primary sources of mortality data are the internationally accepted form of the death certificate, which is used in Yugoslavia. However there are certain limits on its reliability when used in Yugoslavia and these are recognised by the authorities. In many cases only the disease leading directly to death may be recorded, and antecedent causes upon which the international statistical classification is based may not be given. Due to the comparative scarcity of doctors in Macedonia, diagnosis of the cause of death may be made by a person other than a doctor, and the difficulty is further compounded by the lack of medical treatment so that a doctor may be confronted by a cadaver about which little of the medical history is known. In Macedonian published figures of deaths by cause of death, these problems are recognised and information is given as to the number of deceased in each disease category who received treatment prior to death, and also the category, be it doctor, medically qualified layman, or other person who diagnosed the cause of death.

As an example we may cite the 1966 returns of death by cause of death in Macedonia.[13] In that year there were 255 deaths ascribed to influenza. Forty-one had been treated before death, and the cause of death was certified in only six cases by a doctor, seven cases by a qualified layman, and 242 by other persons. Such figures are not, however, typical, although the problem exists to a greater or lesser extent with most disease categories.

Further problems arise when the collected information is published, for in the first place, no information is published as to the sex of the deceased, so that no areal differentiation by sex would normally be possible. Thus where an occupational factor affecting the male population is suspected in the pathology of a particular disease, it is difficult to isolate those who have been involved. Secondly, no indication is given of the ages of those who have died, except where the data appears as part of the national return for the whole of the Republic.

Two final problems exist. There is no separation of rural and urban areas in the published figures, and, further, deaths are only published by the simplified categories of the 'B' list of the I.S.C. (1957). These four difficulties were overcome by personal contact with the Macedonian Statistical Office, which was kind enough to supply the information.

Recognised methods of comparing areas by the use of standardisation techniques must be employed with care in the Yugoslav situation. Using the national age-specific disease rates[14] as a norm, the usual procedure is to apply these rates to the corresponding age group in the local area, thus arriving at a number of expected deaths in that age group in the local area. The age-specific expected deaths in the local area are then summed to give a total number of expected deaths. This figure, divided into the actual deaths which have occurred over the same period, and multiplied by a hundred, gives the standardised mortality ratio.[15] Local S.M.R.'s are therefore directly comparable since they are made always with reference to a national norm or level.*

The mortality maps for Macedonia, which are presented in this paper, have been calculated by this method for the period 1965–7. Such a short period would normally be a minimum period which would give a reliable result, and, other things being equal, the longer the period the more reliable the results would be, except, of course, where a disease shows a temporal cyclical pattern. In Macedonia the boundaries of the local administrative

---

* See also pp. 194–5 for details of calculation.

areas (communes) which are used for the basis of the statistical service were changed three times in the early 1960's. Fortunately most changes have been simple amalgamations of contiguous communes, enabling statistics for the two areas to be easily combined, but that and other difficulties, such as the loss of material in the 1963 Skopje earthquake, preclude use of figures from earlier dates.

Satisfactory mortality data can only be used to best advantage where reliable population figures are also available. The middle year of the figures used here is 1966, which was also the mid-censal year in Yugoslavia, and reliable estimates of total population in each commune were available. Using a simple proportional method, based on the 1961 census, this local estimate was broken down first into its urban and rural components, and then into five-year age groups, by sex.

The complete data of both cause of death by disease, sex and rural and urban parts of the commune, and reliable estimated population figures were thus obtained.

THE MAPS

The base map upon which mortality and morbidity have been plotted is the administrative commune boundaries of Macedonia for 1965. Circles represent the nine major urban areas, and although population centres exist in the other communes they are not considered to be truly urban by reason of their small size, and the nature of their economy. The standardised Mortality Ratios and Standard Morbidity Index are represented as very high, high, normal, low or very low for the sake of simplicity in these examples, although in the case of influenza there is no division between low and very low mortality because many communes have less than one fatality from influenza in the three-year period 1965–7.

Figs. 3.1 and 3.2 illustrate the male and female patterns of mortality from influenza (I.S.C. 480–3) respectively, in the period 1965–7 in Macedonia. Fig. 3.1 shows the higher mortality to be confined almost exclusively to the west of Macedonia, although two communes in the east also have high rates. The urban areas all show low or normal rates, as does that of the central southern region. The female pattern is similar to that of the male with high mortality generally in the west.

For reasons of comparison with the morbidity map of influenza Figs. 3.1 and 3.2 are generalised in Fig. 3.3, the urban areas being considered part of the commune in which they are found, and male and female rates being included together. The map illustrates the major features of the other two.

Fig. 3.4 illustrates the pattern of morbidity for the same disease according to the method of standardisation already discussed. The same base map as used for mortality is used on which to plot morbidity, although as we

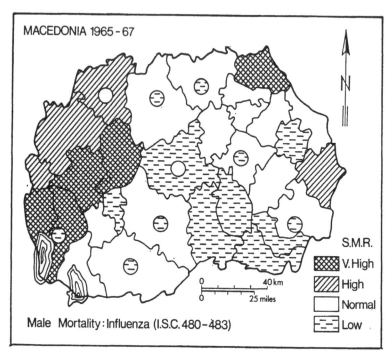

FIGURE 3.1

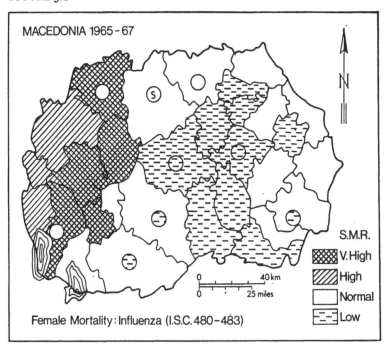

FIGURE 3.2

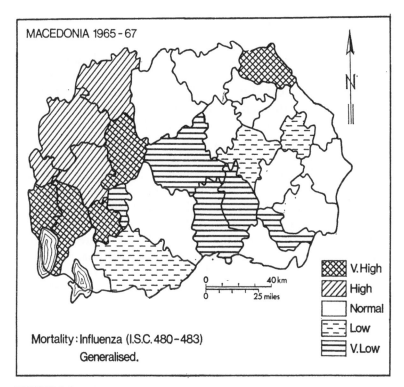

FIGURE 3.3

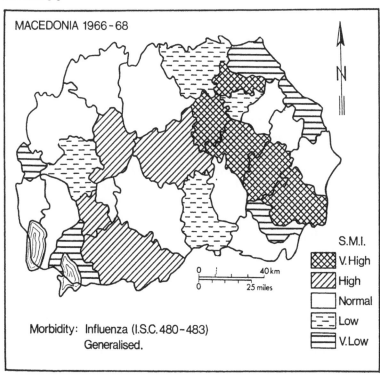

FIGURE 3.4

have seen it may not be the most satisfactory. The distribution illustrated by Fig. 3.4 is quite different to that of the generalised mortality map. In particular the western part of the Republic, which had high mortality, has, with the exception of two communes, only a low morbidity. The area of highest mortality has only a low morbidity. The inference of this would seem to be that, whilst the west had relatively few cases of influenza, using the standardisation technique, there is a higher level of mortality there than elsewhere in the Republic. This suggests that susceptibility to fatal contraction of influenza is, in general, higher in the west than in the east.

Where mortality is low, however, morbidity seems to be higher, suggesting that there is a great difference in the reaction to the influenza virus between the west and the east of the Republic. If this difference was shown to be persistent over a longer period, one might suggest that continued exposure to the disease in the high morbidity areas results in the acquisition of a resistance to the disease, thereby decreasing the number of potential fatalities. This would be only one explanation of the phenomenon. There are others, such as differences in treatment, levels of social and economic development, and environmental differences which are worth considering.

Included in these general headings would be such particular variables as diet, climate, affecting both the human host, and the development of the virus, levels of sanitation, and exposure to the virus in the working environment. The effect of vaccination campaigns in this latter respect should be anticipated.

Fig. 3.5 illustrates the generalised pattern of mortality in Macedonia from the two groups of diseases, arteriosclerotic and degenerative heart disease (I.S.C. 420–2) and chronic rheumatic heart disease (I.S.C. 410–16), by a map produced in the same way as that for influenza. In the more detailed maps of mortality, which are not produced here, the male and female patterns for the two groups of diseases was similar. A generally higher mortality in the east of the Republic could be identified, with the urban centres of the east much higher than the rest, as were the communes surrounding the Republican capital, Skopje. Fig. 3.5, the generalised map, illustrates the major division between the east and the west, although there are two adjacent communes in the east with low mortality.

Fig. 3.6, the morbidity map for this group of diseases, has a much more disparate distribution than its counterpart, the mortality map. Apart from the three communes of very high morbidity in the south-west, other high levels are found, mainly around Skopje, and the very clear delineation found between high and low mortality is not repeated for morbidity. In comparing the two maps there are only two communes where the difference between morbidity and mortality is very significant, i.e. very high and very low. Other communes do differ, but only between high or low and normal. For example, the high mortality level of the south-west is matched by a high

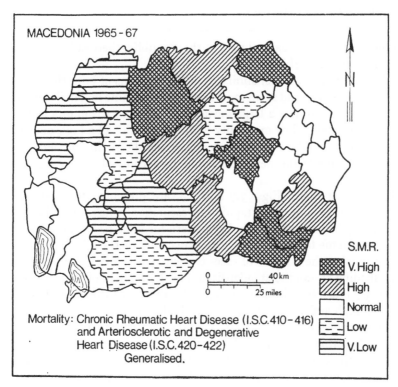

FIGURE 3.5

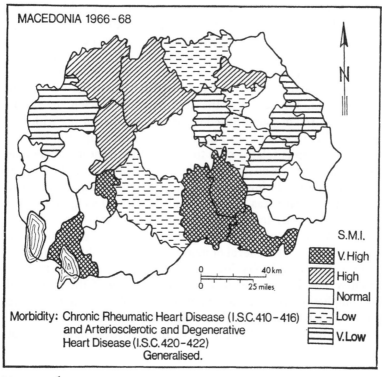

FIGURE 3.6

morbidity, but in the central part of the Republic, there is high mortality but only normal morbidity.

Where there is a general similarity in the patterns of both the mortality and morbidity, it would seem most relevant to examine in detail those communes where there was a large deviation. The comparison of a commune with high mortality/low morbidity with one which had a high morbidity/low mortality might suggest differences significant in the aetiology of the disease, unless the deviation could be accounted for by extraneous factors.

Two further maps, Figs. 3.7 and 3.8, are presented to illustrate two further techniques of cartographic representation which might have been used. Fig. 3.8 shows the morbidity pattern for influenza using a base map derived from the total number of cases treated at each clinic. Each circle is proportional to the number of cases treated at the clinic, whilst the choropleth shading shows the standardised morbidity index of the disease.

Fig. 3.7 illustrates the same material by the use of isopleths. The gradients and high and low areas can be easily determined.

The examples used here, from Yugoslav Macedonia, to illustrate mortality and morbidity mapping, have raised a number of points. In order to fulfil a comparative role, maps of both types were drawn to the same base. The commune base maps used are satisfactory for the portrayal of mortality data, and similar base maps have been used elsewhere. However, they assume a finite boundary which may not be applicable in morbidity mapping, where such boundaries may not exist. The alternative, of which Figs. 3.7 and 3.8 are examples, is to use bases which are not related to areal boundaries or populations. The base map for Fig. 3.8 is a number of circles proportional to the number of patients treated at the clinic, and is derived from demographic base maps used and suggested by others.[16] The isopleth, or isomorb map, Fig. 3.7, has the advantage of showing the morbidity gradient, expressed in standardised terms, and yet dispensing with internal boundaries. The fact that both Figs. 3.7 and 3.8, are superimposed on the administrative map is related only to the need for a means of comparison in this example. A more adequate method would be to draw the maps without reference to the administrative boundaries, providing an overlay to aid location of particular areas. There would then be no suggestion that distributions refer wholly to particular administrative areas.

For the sake of simplicity in the presentation of these maps it will be noted that divisions in the level of mortality and morbidity, into very high, high, normal, low and very low, are made on the basis of rank order. Quantitative divisions would increase the accuracy of the maps without undue complexity, and further allow more sophisticated techniques of statistical analysis to be used when comparing different distributions.

So far no mention has been made of the statistical significance of the

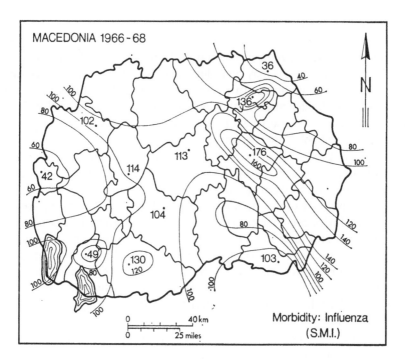

FIGURE 3.7

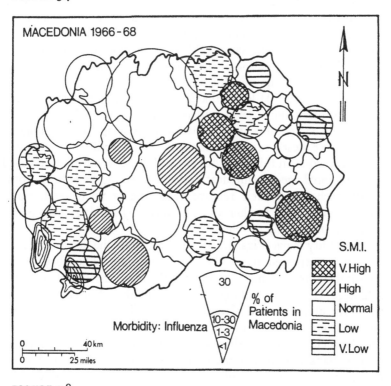

FIGURE 3.8

distributions which have been illustrated. In the case of influenza only the very high mortality areas may be judged to be significant, since large numbers of deaths are involved. In respect of many of the others there were few cases, and the distribution thus obtained may be very doubtful. In the case of the heart disease category the numbers involved were large and significance probably much greater.

Aside from these criticisms, how valuable are such maps as these? Mention has already been made of the part that can be played by the use of mortality and morbidity maps in conjunction, in illustrating those areas where susceptibility to the disease may be fatal. Provided that they are used carefully and with consideration to the factors involved in their production, they can be useful tools in epidemiology, not merely as a convenient form of expressing the areal or spacial characteristics of the disease, but also as pointers to further investigation of the aetiology of the disease. Care must be exercised in the choice of map, be it morbidity or mortality, with reference particularly both to the characteristics of the disease, and to the purpose of the map.

The value of such maps of these could be improved by making a number of minor alterations. It has been suggested that comparability with other diseases would increase the value of the maps, and this may be achieved quite simply using the maps used in this example. It can be seen that there is a wide variation in the level of mortality from the two disease categories in most local areas and the same holds true for morbidity. Other diseases may have closer relationships to each other. International mortality comparisons are simply made by adjusting the age-specific death rate, used in the calculation of the S.M.R., so that it is based, not on a national level, but rather on an international level. Indeed it may be possible to produce an internationally comparable and compatible map using a notional figure for the age specific death rate. No absolute values could be attributed to S.M.R.'s arrived at in this way, but since all S.M.R.'s would be based on the same notional age-specific death rates they would be entirely comparable with each other, their relative values being unchanged. Such an adjustment would add further value to maps by making them part of a much larger view of the problems.

The morbidity maps (Figs. 3.4 and 3.6) illustrate the degree of precision which can be expected from Macedonian sources. The absence of the urban/rural and male/female divisions of the maps are in particular a great disadvantage. However, working within the framework already constructed by the form of the statistical data, it is possible to arrive at estimated statistics, the mapping of which in whatever form can give at least a relative idea of the incidence of the disease, and at best a quantitative assessment of it. Such maps may then be useful in assessing the value of public health schemes, private treatment, and preventive medicine, prediction of future trends of disease, demand for medical services, and also provide some indication of further steps in the investigation of disease.

Such further investigations could include work by many specialists in different disciplines, but might include the investigation of social variables such as family size, overcrowding, diet, ethnic differences, customs and habits, economic factors, including occupational hazards, poverty and malnutrition, or biochemical and biological parameters such as the microclimates affecting disease organisms, and climatic factors affecting man.

In this the geographer who has interested himself in problems of disease distributions and patterns has a peculiar talent which may well be useful. By the nature of his training he has been taught to use information of all descriptions from many and diverse sources. His peculiar skill often results from the synthesis of facts and information, and the coordination of ideas quite unrelated to each other in the normal fields of research. It is this open-mindedness towards information from all sides and its national utilisation which is perhaps in the end his most worthwhile contribution, for he is in a unique position to interpret his maps in terms of their meaning for the possible relationships of the wide variety of factors in the human environment.

REFERENCES

1 MCGLASHAN, N. D. (1966) *Int. Pathology*, **7**, 81.

2 MCGLASHAN, N. D. (1965) *S. Afr. med. J.*, **47**, 35.

3 GELYAKOVA, T. M., VORONOV, A. G., NEFEDOVA, V. B. and SAMOGLOVA, G. S. (1967) *Soviet Geography: Revue and Translation*, **8**, 228–34.

4 MOMIYAMA-SAKAMOTO, M. (1966) *Acta Geologica et Geographica Universitatis Comenianae Geographica, No. 6.*

5 SNOW, J. (1855) *On the Mode of Communication of Cholera.* 2nd Edition. London.

6 MAY, J. M. (1958) *Studies in Medical Geography* (2 vols.). New York, 1958 and 1961.

7 HOWE, M. G. (1963) *National Atlas of Disease Mortality in the United Kingdom.* Royal Geographical Society, London.

8 LEARMONTH, A. T. A. (1961) 'Medical Geography in India'. *Geogrl. J.*, **127**, 10–26.

9 MCGLASHAN, N. D. (1965) *op. cit.*

10 LEARMONTH, A. T. A. and NICHOLS, G. C. (1965) Australian National University, Department of Geography, *Occasional Paper No. 3.*

11 World Health Organisation (1957) *Manual of the International Statistical Classification of Diseases Injuries and Causes of Death* (2 vols.). W.H.O., Geneva.

12 W.H.O. (1957) *op. cit.*

13 Republički Zavod za Statistika (1967) *Prinčini za Smrt vo S.R.M. vo 1966 Godina.* Skopje.

14 The age-specific death rate is normally expressed as the death rate per thousand persons in each age group usually of five years. The national age specific rates are calculated from national returns of deaths, by cause of death, sex and age groups published annually in *Demografska Statistika*, Zavod za Statistika, Belgrade.

15 HOWE, G. M. (1960) 'The geographical distribution of cancer mortality in wales, 1947–53'. *Trans. Pap. Inst. Br. Geogr.*, **28**, 199–214.

16 FORSTER, F. (1966) *Br. J. prev. soc. Med.*, **20,** 165. Reprinted p. 59 *et seq.*

4 Use of a Demographic Base Map for the Presentation of Areal Data in Epidemiology*

F. Forster

The purpose of this article is to offer a development of the demographic map as an alternative to the geographical base-map for the presentation of areal data in epidemiology. The results of geographical investigations into disease morbidity or mortality are often presented cartographically and invariably the base-map used is the normal geographical one, showing the relevant administrative areas. The relating of disease rates to area is useful, in that the recognition of areas characterised by high or low rates may lead to clues of aetiological significance.

When considering areal patterns of disease, however, the epidemiologist requires information about the size of the population at risk in the areas concerned. Sutherland (1962),[1] drew attention to the principal deficiency of the geographical base-map in this respect. Referring to Scotland, he showed that on the normal map correct weighting could not be given to the large urban populations which occupy small areas, whilst small rural populations, often sparsely distributed over large areas, could appear to be over-represented. Thus, base-maps which would relate disease rates to the local populations at risk as well as to geographical position might prove useful epidemiological tools. Development of the demographic map, in which the area of each administrative unit is made proportional to its population whilst contiguity of geographical boundaries and the relative geographical positions of the units are maintained as far as is possible (Hollingsworth, 1966),[2] offers interesting possibilities in this direction.

The use of such maps in epidemiology is not new, Levison and Haddon (1965),[3] for example, having used the technique for plotting cases of Wilm's tumour in New York State. To date, however, these maps have not been used to present data at the national level.

Sutherland (1962) developed a form of demographic diagram, using the same principle of area proportional to population, called an isodemic representation. As a basis he split Scotland into its five hospital regions and

* Reprinted from *British Journal of Preventive and Social Medicine*, (1966) **20**, 156–71.

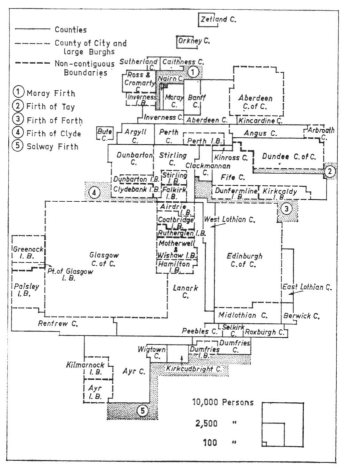

FIGURE 4.1 Locational cartogram: Public Health districts of Scotland. Population base, 1961 census, females aged 45 to 54 years.

showed them separated. He thus lost geographical contiguity between the administrative units, a condition it would seem desirable to preserve.

An attempt has been made in developing the age-sex specific cartograms illustrated (Figs. 4.1, 4.3 and 4.4) to relate disease rates to both the population at risk and to geographical position in the fifty-seven public health districts of Scotland.* A basic aim was also to try to produce cartograms similar to each other in local outline as well as in overall shape to make them readily comparable and hence of greater potential utility. Efforts to

* The insular portions of Inverness and Ross and Cromarty Counties, treated as separate public health districts by the Registrar General for Scotland, have been considered as part of their parent counties here, thus reducing the number of districts from fifty-nine to fifty-seven.

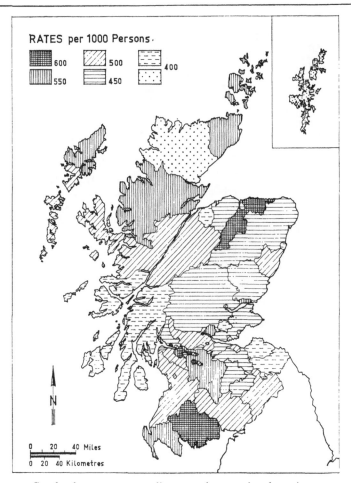

FIGURE 4.2 Scotland: average mortality rates (conventional map).

simplify their construction and final shape have also been made. To date demographic maps have represented complicated construction problems and their final appearance has often been of considerable complexity. These factors may well have obscured their advantages and contributed to their limited application so far.

METHOD

Levison and Haddon (1965) outlined their procedure for constructing a demographic map or, as they called it, an 'area adjusted map' of New York State. As there is no unique solution for a given area and population, and because conventions are developed as one proceeds which require explanation, an account of the method of construction used here is given below.

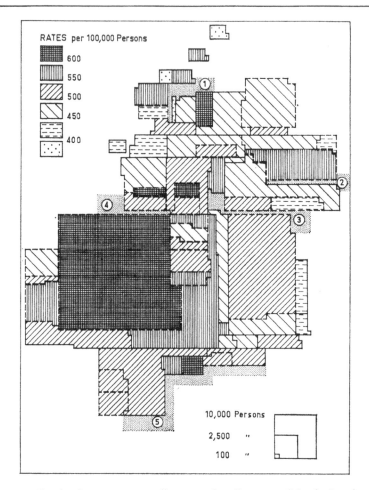

FIGURE 4.3 Scotland: average mortality 1959–63, all causes of death, females aged 45 to 54 years (cartogram).

Reference to the locational cartogram (Fig. 4.1) based on the 1961 female population in the age group 45–54 years, will facilitate understanding of the following explanation:

   (1) 1961 census population figures were taken for the fifty-seven public health districts of Scotland for the 10-year age groups 35–44, 45–54 and 55–64, and for 35–64 combined for males and females separately. There were thus eight age-sex groups in all.

   (2) A scale of 0·1″ square to the nearest 100 persons was selected as suitable for plotting the cartogram outlines on to arithmetic graph paper.

   (3) The firths of Moray, Tay, Forth, Solway and Clyde were selected as

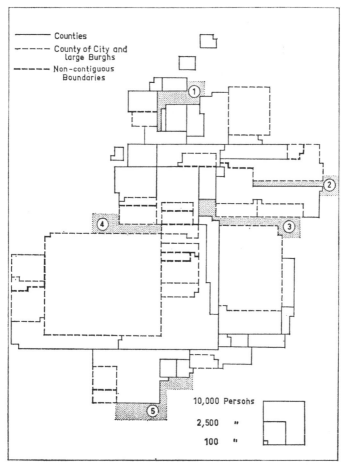

FIGURE 4.4 Scotland: age-sex specific cartogram (outline), population base, 1961 census, males aged 35 to 44 years.

prominent geographical features around which to build the carto-grams. (These features penetrate markedly into Scotland and do much to determine the country's shape.)

(4) The cities of Glasgow, Edinburgh, Aberdeen and Dundee were given constant positions fronting on to their respective firths. As these cities are important population concentrations, their positions and size partially determined the positions of the other units. The firths and cities thus provided a framework of lines and areas common to each cartogram which helped to achieve similarity of final outline.

(5) To prevent the cartograms from becoming too complex in shape, a factor which can lead to difficulties of assessing the relative weighting

to be given to individual units, the following conventions were adopted:

(a) No attempt was made to retain the geographical shape of administrative units; instead they were formalised into shapes based on rectangles. Consideration of Argyll with its small population, considerable north–south extent and often insular shape will show why this was necessary. Any attempt to retain its shape on a demographic map would have resulted in a fragmented narrow zone, the demographic weighting of which would have been difficult to assess.

(b) Geographical positions were modified where it was felt that this helped towards simplification of shape and construction, whilst as far as possible geographical contiguity of boundaries was maintained. For example, relative to Ayrshire, Wigtown has been given a north-westerly, as opposed to its true south-westerly, position.

(c) Where two or more large burghs occurred in a county, they were grouped and indicated as non-contiguous. The alternative, to have distributed them throughout their county area, would have fragmented the latter and again have contributed to the problem of assessing demographic weighting. In reality, no two large burghs in any county are more than 15 miles apart and in Lanark, which has five large burghs, both Airdrie and Coatbridge, and Motherwell and Hamilton are adjacent. Grouping does not therefore cause such a distortion as it might at first appear.

RESULTS

The same mortality data, calculated for all causes of death among females aged 45–54 years for the period 1959–63, is presented on the conventional map (Fig. 4.2) and the cartogram (Fig. 4.3). Clearly, impressions gained from maps are subjective and opinions will vary as to the relative value of different forms of presentation. Attention is drawn below to only the major points of difference between the two methods shown here in an attempt to illustrate the characteristics and, it is felt, the essential advantage of the demographic map.

As a first step it will be helpful to compare the locational cartogram (Fig. 4.1) with the geographical map (Fig. 4.2) so that familiarity can be gained with the conventions of the former and the alterations in the areal weighting of Scotland that result. In particular (Fig. 4.1):

(1) The north-west highland counties, north of Dumbarton and Stirling Counties, and the southern ones from Berwickshire to Wigtown, with their numerically small populations, are reduced in area, relative to Scotland as a whole.

(2) Conversely, the main areas of population concentration in west central Scotland and in eastern Scotland from Edinburgh northwards assume increased proportions.

(3) As a result of these changes, it is easier to appreciate the extent to which the population group used as a base for the cartogram is resident in the larger urban centres (Table 4.1) and in Glasgow in particular.

TABLE 4.1 *Selected Population Totals, 1961 Census, Females aged 45–54 Years*

Area	Population	Per cent of Scottish Age Group
Glasgow	72,885	20·44
Edinburgh	35,683	10.00
Aberdeen	13,619	3·82
Dundee	12,896	3·60
Counties of City	135,083	37·86
Large Burghs	50,317	14·11
Cities and Large Burghs	185,400	51·97
County Areas	171,295	48·03
TOTAL	356,695	100

Comparison of the geographical distribution of the mortality rates on the two presentations (Fig. 4.2 and Fig. 4.3) can usefully be made at two levels. At the local level it is interesting to note the way in which the following are represented:

(1) Glasgow County of City and Kirkcudbright County, both being high mortality rate areas.
(2) Dundee County of City and Angus County, respectively high rate and low rate areas.
(3) The variable mortality experience in the large burghs in geographical proximity to Glasgow.

At the wider regional level, the following summary of the geographical pattern of mortality is obtained and, it is suggested, is easily recognisable on both presentations.

(1) Glasgow is the most prominent area of unfavourable mortality experience. Together with the large burghs of Stirling, Airdrie, Hamilton, Paisley, and Lanark County, it forms a high rate area in west central Scotland. Within this area the large burghs of Greenock and Coatbridge stand out in marked contrast by virtue of their low mortality rates.

(2) No other concentrations of high mortality are found. Rather, a sporadic distribution of high rate areas occurs, including the rural counties of Orkney, Caithness, Banff, and Ross and Cromarty in the north, Wigtown and Kirkcudbright in the south, and the urban area of Dundee on the east coast.

(3) A zone of moderate to low mortality experience extends from Argyll and Bute, north-eastwards to Aberdeen County and south-eastwards to Berwickshire.

(4) All of eastern Scotland south of Aberdeen County, with the exception of Dundee, is thus a favourable area in terms of mortality experience in this sex and age group.

DISCUSSION

The principal value of a map is that it enables the essential features of a distribution pattern to be quickly appreciated. The cartographic needs of different disciplines do, however, vary, and maps are of greatest value when designed to serve specific requirements. The cartograms presented above are illustrative devices especially suited to the needs of the epidemiologist working with areal data for, by their very nature, they correct the principal defect of the conventional map by illustrating the required dimension of population.

Perhaps the most encouraging feature is that, as a result of careful initial consideration of the major demographic and geographic features of Scotland, it has been possible to produce a demographic base-map which is simple enough in arrangement readily to allow geographic patterns to emerge. Further, the utility of the device is considerably increased by the fact that the eight different age-sex cartograms produced are easily visually comparable and that they all share a common set of conventions (cf. Figs. 4.1 and 4.4). It will be clear that such maps can be prepared for any area and any subsection of the total population, given the relevant population figures. It might be useful to consider the production of a national series of age-sex specific base-maps of this type having common shapes and conventions, for use in the presentation of areal data in epidemiological studies.

SUMMARY

Presenting disease rates on the conventional geographical base-map does not allow of weighting for local population differences. Development of the demographic map offers possibilities for relating disease rates both to local populations at risk and to geographical position. In an attempt to achieve these effects, age-sex specific cartograms were developed for Scotland based on 1961 census figures. Their construction is described and, by presenting the same mortality data on both the cartogram and the geographic base-map, an attempt is made to indicate the advantage of the former over the latter for the presentation of areal data in epidemiology.

ACKNOWLEDGEMENTS

I am indebted to the following: Dr. Mary Fulton and Mrs. Elspeth Semple for their kind permission to use the epidemiological data portrayed in this article, prof. S. L. Morrison for his encouragement, Mr. T. Edge of J. Bartholomew and Company Limited for advice regarding the use of cartographic equipment, and Prof. Wreford Watson for kindly placing at my disposal the facilities of the Department of Geography in the University of Edinburgh.

REFERENCES

1 SUTHERLAND, I. N. (1962) *Br. J. prev. soc. Med.*, **16**, 30.
2 HOLLINGSWORTH, T. H. (1966) Map of Results of General Election. *The Times*, April 4, 1966. Late London Edition, p. 8.
3 LEVISON, M. E. and HADDON, W. (1965) *Publ. Hlth Rep. (Wash.)*, **80**, 55.

5 Computers and Mapping in Medical Geography

R. W. Armstrong

One of the more promising developments for medical geography during the 1960's was the application of the computer to the preparation of data for mapping, and for constructing 'maps' or computer graphics. During this largely experimental period, various modes of output were tried and a number of practical applications demonstrated. The advantages of the technique are essentially those of the computer itself, namely, speed and reliability in handling large amounts of data requiring repetitious analysis. This entails efficient use of the memory of the machine in order to maximise benefit over cost. Large operations requiring considerable data analysis and extensive output of maps, which would have presented a formidable task using desk calculators and draughting pens, can be quickly and economically performed by computer. On the other hand, small mapping projects can prove highly expensive in preparation time and machine costs and are better handled with conventional equipment. The computer must be used efficiently, as a data memory and computing device, and not simply as a printer, if its advantages are to be really exploited in making maps.

COMPUTER GRAPHICS

Computer graphics is a cover term for all kinds of maps or diagrams produced by an electronic computer or its auxiliary equipment.[1] Most cartographers prefer not to call the computer graphic a map because it is mechanically drawn. The finished product consists essentially of lines or point symbols located according to a system of coordinates. For example, using the Cartesian system of coordinates, a plotting programme directs the computer to place symbols or draw lines within a grid. By giving the computer an x and y coordinate and instructing it to use a particular symbol, that symbol will be positioned precisely on the grid. In Fig. 5.1 a simple plot is illustrated where characters have been placed by a lineprinter within a grid measuring 100 character spaces along the horizontal x axis, and 50 up the vertical y axis. This is a convenient size for one sheet of standard printout paper. Other modes of map output include plotting machines and cathode ray displays.

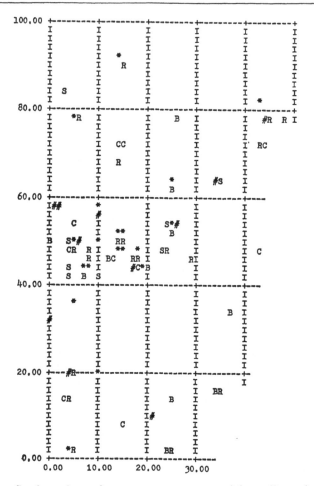

FIGURE 5.1 Section of a point-type map constructed by a line-printer which positions symbols on a grid according to predetermined coordinates. This simple computer graphic shows the location of infant deaths and a number of environmental health hazards reported to a city health department. Symbols are: * infant death, R refuse violation, B burning rubbish violation, S sewage hazard, C abandoned vehicle, and # junk or timber pile constituting rat harbourage.

The data input required to produce the simple line-printer plot in Fig. 5.1 consists only of a set of instructions for symbols and a set of co-ordinates for locating the appropriate symbols on a grid. The primary input media are usually cards or paper tape. The programme, or set of instructions, required to perform the operation in Fig. 5.1 is comparatively simple and requires little computer memory capacity. It can be performed efficiently on accounting machines.[2] Where large amounts of data are to be used, or where analyses are to be performed on the data in preparation for mapping, machines of large memory capacity are necessary. A good example of the application of computer graphics to a project dealing with large amounts of data is the *Atlas of British Flora*.[3]

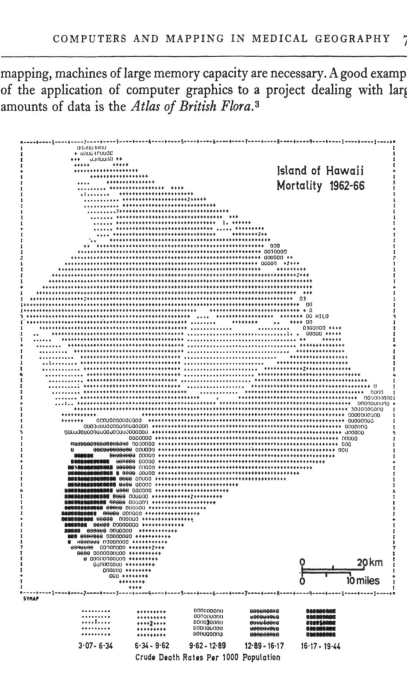

FIGURE 5.2 Statistical map of mortality of the Island of Hawaii using 20 data points corresponding to death rates for census tracts. Isopleths were computed mathematically from the data points and according to five levels made up with an equal proportion of the value range of the data in each level, i.e. 20 per cent. This computer graphic was constructed using the Synagraphic Mapping Program (SYMAP), Version 5, of the Laboratory for Computer Graphics and Spatial Analysis, Harvard University, Cambridge, Mass.

Line-Printers

The output form of maps in medical geography has been largely restricted to line-printers, a reflection no doubt of their convenience and comparative low cost. But they leave much to be desired in the way of appearance. The fixed symbol size places restrictions on the scale of the map and there are some limitations due to the symbols themselves. It is possible to instruct the machine to overlay successive symbols so as to build up a gradation of shadings from white (blank) to black (Figs. 5.2 and 5.3). Lines can be

TABLE 5.1 *Data for Figs. 5.2 and 5.3*

Hawaii Census Tract Number (1960 Census)	Crude Death Rate Per 1000 Population 1962–1966	Fig. 5.2 (Five Mapping Categories) Level Assigned	Fig. 5.3 (Six Mapping Categories) Level Assigned
1	9·68	3	5
2	8·37	2	4
3–4–5	9·98	3	6
6	3·93	1	1
7	10·21	3	6
8	7·30	2	3
9	8·49	2	4
10	11·93	3	6
11	6·77	2	3
12	6·15	1	2
13	8·34	2	4
14	10·86	3	6
15	3·07	1	1
16	6·44	2	3
17	9·44	2	5
18	5·26	1	1
19	6·15	1	2
20	19·44	5	6
21	8·49	2	4
22	7·29	2	3

Mean death rate for County, 7·64 per 1000 resident civilian population, 1962–6; standard deviation of rates, 3·36. Population and mortality data supplied by Hawaii State Department of Health.

simulated by rows of a certain symbol, but the drawing of curves is obviously awkward and unless the scale is very large detail is lost. Lines may also be indicated by leaving the strips between blocked-in areas blank (Figs. 5.2 and 5.3). The capacity of the line-printer as a maker of maps has been thoroughly explored by the Laboratory for Computer Graphics at

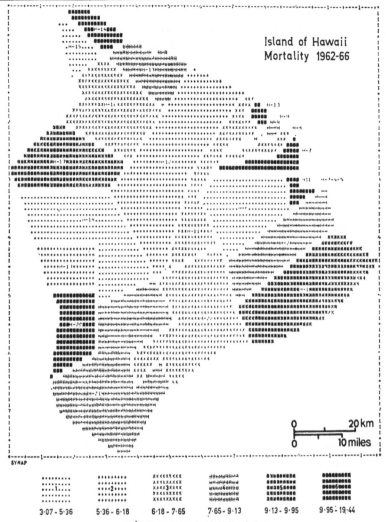

Island of Hawaii
Mortality 1962-66

0   20 km
0   10 miles

SYMAP

3·07 - 5·36   5·36 - 6·18   6·18 - 7·65   7·65 - 9·13   9·13 - 9·95   9·95 - 19·44

Crude Death Rates Per 1000 Population

FIGURE 5.3 Statistical map of mortality of the Island of Hawaii using the same
mortality rates as in Fig. 5.2 but plotted by a line-printer as a choropleth map with
census tracts as units. The six levels of symbolisation were predetermined using
the standard deviation of the distribution of rates and are equivalent to equal areas
of the standard normal curve. This computer graphic was constructed using the
Synagraphic Mapping Program (SYMAP), Version 5 of the Laboratory for Computer Graphics and Spatial Analysis, Harvard University, Cambridge, Mass.

Harvard University in the development of its SYMAP series of programmes.[4] In terms of mapping, the line-printer can produce good point
and area symbols, for point- and choropleth-type maps respectively, but
it is not so useful for isopleth maps.

Plotting Machines

Plotting machines (such as the CalComp Plotter) would appear to be ideal for drawing isopleth or contour maps. They are usually of the digital incremental line variety, which plot short straight lines according to a set of x, y coordinates. The resulting line appears fine and smooth and any colour of ballpoint pen may be used. However, for any detailed isopleth operation a large computer memory is required in order to store the matrix of control values through which the programme traces, interpolating to find the points through which the isopleths pass. Plotters are normally 'off-line', i.e. not run as an integral part of the main computer, and thus require separate drive equipment, all of which adds to costs. Perhaps the greatest drawback at present for isopleth mapping is the difficulty of programming, especially where a small number of data values are being used. Results tend to differ considerably from those created manually by a cartographer. Further experimentation will no doubt lead to plotting machines being more generally useful, and hopefully, some reduction in costs can also be managed.

Cathode Ray Display

This form of output offers advantages of speed and flexibility. In situations where direct communication with the computer is available, the output as viewed on the cathode ray display can be quickly changed if required by giving new data or analytical instructions to the machine. A permanent record of the display can be made if desired by photography. However, this mode of output does not serve the needs of geographers as well as those modes using paper recording. Unless the cathode ray tube is of large dimensions and a fine line-scanning system is employed, the output results on photographs are poor. The various kinds of output for computer graphics have been reviewed by Hopps and others who provide several examples.[5]

DATA PROCESSING

The data processing necessary to convert raw data into a finished computer graphic can be considered as four phases: (1) the *inputs*, which convey data to the machine, (2) *analyses*, which prepare the data for presentation, (3) *presentation*, or the method used to convert analysed data into map form, and (4) the *output* mode.

Input Phase

The inputs, in the form of punch cards or tape, comprise the programme, or programmes, instructing the machine to perform all required operations. In computer graphics the inputs also contain the coordinates for establishing map outlines such as for coastlines or county boundaries, the co-ordinates for the location of data points or symbols representing data, and the data values themselves. There are now available for purchase a number of excellent package programmes for computer graphics which can be stored in a computer library. To call such programmes into use usually requires only one or two callcards in the input deck.

Analysis Phase

It is in the analysis phase that the computer can be used to greatest advantage in the compilation of medical maps. For example, the preparation of a statistical map depicting mortality rates by a number of geographical units, such as counties, requires a large number of tedious computations which can be quickly and accurately performed by the machine. In the computation of age-adjusted rates (indirect method) or standardised mortality ratios, the input of data values would consist of the age-specific probabilities of death for the total population, fed in as constants, and for each geographical unit, the unit populations by age-groups and the actual number of deaths. Because the computer is performing the labour of calculation, precision can be improved by increasing the number of age-groups conventionally used and by carrying all working to several decimal places.[6] In addition, it is usually of advantage to process data for a number of different maps in the same operation. For instance, statistics could be prepared for maps of rates by different age-groups, causes of death, or time periods.

Computers are particularly useful for storing, sorting, comparing and classifying large amounts of data. Data stored on cards, tape or disk can be fed to the machine when study demands. Comparable data on a series of medical and environmental variables, for example, each given geographical location by coordinates, can be stored and compared ready to print out a series of maps depicting their distributions. It is comparatively simple to add new data to these data storage systems and prepare fresh maps that regularly monitor changing geographical patterns.

Presentation Phase

The presentation phase is that set of operations concerned with preparing the data for the kind of map desired, i.e. point, choropleth, or isopleth. This takes the form of a set of programme statements concerned with establishing the grid system on which data, or their representative symbols, will be arranged for printing, and any necessary analyses such as those required to produce an isopleth map. Package programmes, such as SYMAP,[4] usually allow a variety of choices of map presentation. SYMAP permits three main types: contour (isopleth), conformant (choropleth, conforming to predefined areal units) and proximal (choropleth, but where the areal units are established by nearest neighbour methods from data points).

Point Maps The simplest kind of presentation is to instruct the line-printer to plot symbols at coordinate locations on a grid (Fig. 5.1). This procedure was described above. If desired, the printer can include a line of dots, or x's, etc., roughly to indicate a coastline, river, or other feature, and print any labels and titles. In Fig. 5.1, qualitative symbols were used to identify environmental health hazards. The programme used to generate Fig. 5.1 places symbols at their respective coordinate locations, but in the case of two or more symbols having the same location, it proceeds to place the second immediately to the right of the first, the third to the left, the fourth on top, and so on in cyclic fashion. A ranked set of values, such as standardised mortality ratios, can be conveniently depicted with numerals, say 0 to 6, representing lowest through highest groups of ratios. If appropriate, actual numerical values can be printed, but there may be risk of crowding.

The choice of symbols suitable for point maps is limited because the printer tends to produce rather uniform characters that are sometimes difficult to distinguish on the resulting map. The dash, letter I and plus sign are usually employed to reproduce the grid, if this is desired. The letter X gives the central point most precisely; the asterisk, zero and letters B, S and Z provide particularly clear contrasting symbols.

Choropleth Maps The choropleth map may be produced on a line-printer using a particular symbol to block in unit areas (Fig. 5.3). Coordinates are given for the boundary corners of all unit areas, which must usually be generalised to have simple straight line borders. The machine is then programmed to fill in the areas with symbols according to the value they are to represent. The SYMAP programme used to make Fig. 5.3 instructs the line-printer to overprint successive characters so as to create a gradation of shadings – up to ten if desired. This kind of map is really a more intricate form of plotting.

Isopleth Maps An isopleth map requires a programme which will trace through a matrix of data values and their coordinates, and interpolate new coordinates through which the lines for mapping will pass. As with manually constructed isopleth maps, the task is easier if there is a high density of data points. But whereas a cartographer can make judgment interpolations by eye, the machine has to be programmed to do the interpolation in some mathematical fashion and may not always yield realistic results. There are several methods in use. Four of the more popular methods are outlined below.

The SYMAP programme sets up data points on a grid and interpolates isopleths using the actual values of the data. A fixed number of isopleth intervals is selected (2 to 10) and the data values are grouped by the programme accordingly. The programme proceeds to enclose the various groups of data points with the appropriate level of isopleth according to their spatial contiguity on the grid. The distance between groups, which positions the isopleth, is established by measuring the half-way point between pairs of values belonging to different groups (in terms of data value, not distance between data points). The result is printed on a line-printer as a series of shadings with the isopleths revealed as the blank space between shadings (Fig. 5.2).

While having the advantage of comparative simplicity, the technique employed by SYMAP is, of course, flexible only within the limits of the programme. Resultant map patterns depend heavily on the density of data points and on the number of isopleth intervals, which is arbitrarily selected by the user. Areas of high or low values may become exaggerated out of all true proportion and in using this technique a variety of isopleth intervals should be tried to seek best results.

An alternative method, one which has been widely used in geology, computes trend surfaces between data point values on a grid using polynomials.[7] One such programme, which is readily available, uses a sixth-degree polynomial surface to print trend surfaces as isopleths on a line-printer.[8] This technique gives good results with a high density of data points but it is complicated to compute.

A line-printer isopleth programme which produces good results with a low density of data points as well as a high density has been developed by Tobler and others at the Department of Geography, University of Michigan.[9] It uses a square grid of varying mesh depending on the number of data points. Grid point values are determined in one of two ways from the data point values. If the data point is within a set distance of the grid point, the grid point receives the value of the data point; if beyond the set distance the value of the grid point is established by a weighted average between the six closest data points. The grid point values form the basis for positioning isopleths.

A more sophisticated technique, using a grid, has been devised by the Control Data Corporation for use with incremental line-plotting machines. This also offers the option of improving the accuracy of isopleths by computing dummy point values by linear interpolation from the original point values. Points can thus be added at levels where original data did not exist. This technique is only available at certain major computing centres.

Other Presentations Other forms of computer graphic presentations have not as yet seen much application in medical geography. The block diagram offers some possibilities especially where the vertical component is important, as in relationships between vectors and disease where the vector and/or disease agent are limited in range by altitude. It could also be of use in preparing geographical models of ecological relationships. Many examples of block diagrams generated by computer have been published.[10] Most depict densities of single variables such as population.

Aerial photographs, either by conventional aircraft or satellite, are being scanned by pattern-recognition computers to collect information for meteorology, geology, oceanography and other sciences. The output can be in the form of a computer graphic either by plot or on cathode ray tube. As yet, there are few direct applications for this technique in human medical problems. It has been shown to have use in detecting thermal pollution of streams and lakes and the distribution of certain plant diseases, and no doubt other applications will appear as experiments continue.

Output Phase

The final output phase has already been referred to in terms of the set of modes, i.e. printer, plotter, and cathode ray tube display. It is usual to print out data analysis results and arrays of computed rates, etc., as well as the map itself. Multiple copies of a map may be made at one time. In the case of point maps, it may be preferred to print out only data values, omitting the grid, coastlines, boundaries and other details. Subsequently, a transparent outline of the coastline, administrative units, etc., can be placed over the computer printout to create the final map. This can then be photographed, or copies made by reproducing machine if a permanent record is desired.

The width of printout paper and its length between folds create some mechanical questions of scale, as does the size of the cathode ray tube. On a line-printer, width is controlled by the carriage, which on most current machines has 120–132 spaces covering about 12–13 inches. Maps covering large areas can be constructed by directing the computer to plot in sections, and afterwards assembling by hand the strips of printout paper to make the large map. Length is virtually unlimited, for the paper is either folded (on line-printers) or on rolls (plotters). While extension in scale is not then a serious difficulty, the smallest scale is limited on a line-printer by the size of character. The smallest geographical mapping unit is the area of the character itself.

Most line-printers print characters at a spacing of six lines per inch. Some print at eight lines per inch, and others can be set to print at both spacings. The appearance of a map employing shading symbols is usually improved if the spacing is at eight lines per inch because darker shadings appear denser and contrasts with lighter shadings are thereby enhanced.

OTHER DATA CONSIDERATIONS

Coordinates

It goes without saying that data intended for mapping by computer will need to use some system of coordinates for location. For most medical mapping purposes Cartesian rather than polar coordinate systems seem more useful. The decimal system of computation in the computer arithmetic favours the use of a metric unit of measurement for coordinates. The military metre grids of eastings and northings, which are indicated on most topographic maps, provide a useful practical system. At scales round 1:20,000 data can be located accurately within a 100 metre square. Latitude and longitude, and other non-decimal systems, require conversion to decimal form. The coordinates for each piece of data can be found from maps and then punched into data cards.

Data Comparability

Mapping of any kind demands close scrutiny of data quality and comparability, but it becomes even more important in computer graphics where the machine produces the same result in outward appearance whether the data is good or bad. There is a tendency to 'believe' the results of a computing machine and it is important to remember that computers do nothing to improve data quality. It is just as easy to be fascinated by an inaccurate computer graphic as by an accurate one!

In comparative mapping, such as where a series of overlays depicting medical and environmental variables are being drawn from a data bank, it is obviously important to be using comparable data in all respects of location, time and measurement which are appropriate to the research problem. One of the major difficulties in creating a data bank drawn from a variety of secondary sources is that the data, collected originally for purposes specific to some other problem, are not directly comparable when brought together. There are many ways of manipulating such data in order to achieve valid comparisons and they should be employed where the alternative of collecting primary data is not possible.

Data measurement characteristics should also be comparable and they should be considered in deciding the form of map presentation. Data should be reduced to the lowest common level of measurement – ratio, interval, ordinal or nominal. Discrete data, such as numbers of deaths, are more appropriately treated by point or choropleth mapping; continuous data, such as temperature readings, are better represented by isopleth. Isopleths may also be preferred in situations where data are scant or unreliable and trends are all that can be indicated. Isopleths were used, for example, in preparing medical statistical maps of Australia.[11]

Mapping Categories

Just as with conventional maps, the number of categories and the distance between categories chosen for symbolisation or shading will greatly influence the nature of the final map and the impressions created for the reader. In computer graphics where several statistical maps of comparable data (such as causes of death) are frequently prepared in a series, it is convenient to use a standard form of categorisation. One method is to use the standard deviation of the array of rates, computed for the geographical units of each map, to establish six categories representing equal areas of the standard normal curve.[12] In this technique all maps are treated in the same way and are directly comparable. The computer can be programmed to perform this standard category operation as one of the steps in the preparation phase of generating a map. The categories in Fig. 5.3 were established using this technique, but not in one operation as the SYMAP programme only permits categories to be established by proportions. The proportions have to be found separately and then fed in as a programme elective.

If a standard form of categorisation is not used, it is preferable to terminate computer operations at the end of the analysis phase and have the rates or ratios to be mapped printed out as arrays. Each array can then be scrutinised and the appropriate number of categories, and distances between them, established for each map. The computer graphics can then be completed using the pre-defined categories.

Time and Cost

Computer graphics can be a very time-consuming and expensive business and before embarking on a mapping project one should always consider the alternative of a piece of paper and a box of coloured pencils! It is very rare indeed to be able to prepare a computer graphic as desired with a single machine run. Even with easy-to-use library programmes, such as SYMAP, there will always be some adjustment of input, especially in coordinates for choropleth mapping, before the result is satisfactory. This involves considerable time for the user and expensive machine time. It is obviously a waste to produce an expensive computer graphic which could have been drawn by hand in fifteen minutes in the office, and often with more pleasing results. Computer mapping becomes attractive where extensive data analyses are required and where many maps in a series are necessary. It is outstanding when used as a mapping surveillance system, where coordinates and programme remain unchanged and only new data values need be added to produce an up-dated map.

APPLICATIONS

While computers can map most topics that can be done by conventional techniques, there are certain applications for which they are better suited for reasons mentioned above. Figs. 5.1, 5.2 and 5.3 illustrate two kinds of computer graphic applied in medical geography.

Environmental Health Mapping

In Fig. 5.1 a portion of a map of a small town is presented showing the occurrence of infant deaths, together with certain environmental health hazards. Overlays of street pattern, housing condition, income level, education, vegetation, prevailing wind direction, and so on, can be placed over the basic grid. The geographical association of infant deaths and environmental health hazards is evident from the computer graphic. These data were collected by sanitary inspectors and public health nurses as part of their routine statistical reporting, but with the addition of geographical coordinates giving the location of each infant death and environmental health hazard. The coordinates were obtained from a large scale map with grid overlay. The system just described has been working in several health departments in the United States and is in the process of being introduced statewide in Oregon. Use of the computer facilitates rapid production of maps on a routine basis with a minimum of fresh data input.

Statistical Mapping

Computer graphics is an excellent way of preparing mortality and morbidity distribution maps because the calculation of rates and ratios can be performed at the same time. They make good working maps for later draughting, a procedure which was followed to prepare maps for the medical section of the *U.S National Atlas*,[13] and which McGlashan and Bond[14] used in preparing fifty-five maps including diabetes mellitus in central Africa using five rankings of reported case numbers. A good example of a choropleth map using standardised mortality ratios has been presented by Howe.[15]

Figs. 5.2 and 5.3 present the same mortality rates by twenty census tract divisions for the Island of Hawaii in order to illustrate isopleth and choropleth computer graphic techniques. It is appropriate at this point to compare these two statistical maps in order to summarise some of the technical considerations which must be considered in making interpretations. Fig. 5.2 uses the standard SYMAP procedure of creating five levels or categories of symbolisation. The mortality rates are sorted into the five levels on the basis of an equal percentage of the absolute value range, i.e. 20 per cent of the range of the rate values fall in each of the five levels. It happened in this particular instance that only one rate was allocated to the fifth (highest) level, none to the fourth, five to the third, nine to the second, and five to the first (Table 5.1). Considerable emphasis is therefore given to the high mortality rate in the south-western region of the island.

In Fig. 5.3, six levels were established, using the procedure outlined above, where equal areas of the standard normal curve are found with the relation: Mean $\pm z$ (Standard Deviation), where $z = 0.431$ and 0.967 to set up six areas of 16.6 per cent in each. For this particular map, five rates were allocated to the sixth (highest) level, two to the fifth, four to the fourth, four to the third, two to the second and three to the third (Table 5.1). If desired, the rates can be interpreted in terms of significant deviation from the mean. The highest rate in this case is more than two standard deviations from the mean, and the lowest more than one. The mapped patterns in Figs. 5.2 and 5.3 give quite different impressions, and interpretations must be very carefully made with reference to the method of allocating shading symbols. Fig. 5.2 emphasises the highest rate in relation to the others; Fig. 5.3 gives more weight to the dispersion pattern of the rates as a group. Neither is superior because both could serve a useful purpose, as long as valid interpretations are provided.

The choropleth mapping technique confines each level to the area representing the census tract, which is, of course, open to misinterpretations in itself. Differences in the area of unit and in the density and spread of the population within units are not accounted for in the choropleth technique. To a certain extent these factors can be incorporated in the SYMAP isopleth technique by using coordinates for data points which are located at the centre of gravity of population distribution, and not merely in the centre of an administrative unit. Crude mortality rates were used in constructing Figs. 5.2 and 5.3. Rates adjusted for age, sex and race would produce different map patterns. These kinds of problems of interpretation of medical statistical maps due to technical considerations have been discussed elsewhere.[16]

Geographical Correlations

Possibly the most exciting prospect afforded by computer graphics is their use as a health surveillance system. In theory, data could be stored and continuously updated and edited for recall to produce maps depicting health (or disease) situations for any part of the world. Other maps of the same areas could be produced showing any desired combination of other variables. By overlay, or statistical tests of association, the degree of geographical association between health related and other variables could be shown and directions for further research indicated. This application of computer graphics was extensively investigated by Hopps and others in a project called 'Mapping of Disease'. The goal was: 'to provide a system whereby disease *and* environmental data could be manipulated together in an appropriate (geographic) location and time context, with direct computer (line-printer or plotter) output in the form of distribution maps, or block diagrams – with supplemental reports as required'.[17] A summary by Hopps of the project report has appeared elsewhere.[18]

Unfortunately, the 'Mapping of Disease' project was terminated before it could become operational. It did, however, demonstrate that the project was feasible. The major difficulties were not technical but centred round those well-known limitations of data availability, quality and comparability. Nevertheless, good progress was made in developing suitable methods for screening data characteristics, in structuring data for storage in the computer, and in developing factor catalogues and data extraction forms. It is to be hoped that this project, or one similar to it, will carry this application of computer graphics in medical geography to fruition.

Service Area Studies

Computer graphics have been widely used to prepare maps in hospital service and ambulance service studies. Drosness and others[19] used plotting programmes used in urban renewal studies[20] to examine hospital utilisation in the Greater Los Angeles Area. They used a computer to prepare maps showing, by census tract, the admissions to each of 127 hospitals during 1962. By comparing admissions to a particular hospital in relation to total admissions to all hospitals, they were able to assess quickly relative hospital utilisation and service in terms of geographic area. Similar applications of computer graphics have been made more recently in a study of hospital utilisation in the Chicago area.[21] Computer graphics lend themselves to the analysis and preparation of all kinds of medical service data where geographic relations are important, as in patient-origin, and time-distance for ambulance and emergency services. The results are of value in planning and research as well as in evaluating current operations.

SUMMARY

Computers offer much as tools which can promote the utility of maps in medical geography. Their speed, data capacity, and accuracy in performing operations make them ideally suited to extensive or continuous mapping operations. The techniques which are now available make possible a versatile selection of map presentations suitable for a variety of studies. The applications are seemingly unlimited. Geographic location, long neglected in most health fields outside epidemiology, is receiving new attention from medical ecologists, health administrators and health planners. Maps are basic models of the spatial aspects of environmental relations and as such have the potential of being basic tools in predictive epidemiology, communicable disease surveillance, and health planning. It is one task of medical geography to develop such maps and to do so with speed and efficiency. Computer graphics provides answers to the latter criterion.

REFERENCES

1 SUTHERLAND, I. E. (1966) 'Computer graphics'. *Datamation*, **12**, 22–7.
2 SOPER, J. H. (1964) 'Mapping the distribution of plants by machine'. *Can. J. Bot.*, **42**, 1087–1100.
3 PERRING, F. H. and WALTERS, S. M., eds. (1962) *Atlas of the British Flora.* Botanical Society of the British Isles: London.
4 FISHER, H. T. *et al.* (1967) *Introduction to Synagraphic Computer Mapping – Computer Mapping of Quantitative and Qualitative Information.* Introductory Correspondence Course, Laboratory for Computer Graphics, Harvard University, Cambridge, Massachusetts.
5 HOPPS, H. C. *et al.* (1968) *The Mapping of Disease (MOD) Project.* Joint Report of The Universities Associated for Research and Education in Pathology and The Armed Forces Institute of Pathology, Washington, D.C.
6 ARMSTRONG, R. W. (1965) 'Graficado mediante computadora en la geográfia médica'. In: Unión Geográfica Internacional (IGU). Conferencia Regional Latino-americana. *Reunión Especial de la Comisión de Geografía Médica* (Mexico) **6**, 69–75.
7 KRUMBEIN, W. C. (1959) 'Trend surface analysis of contour-type maps with irregular control-point spacing'. *J. geophys. Res.*, **64**, 823–34.
8 O'LEARY, M., LIPPERT, R. H. and SPITZ, O. T. (1966) *FORTRAN IV and MAP Program for Computation and Plotting of Trend Surfaces for Degrees 1 through 6.* State Geological Survey of Kansas, Computer Contribution 3, Lawrence, Kansas.
9 TOBLER, W. R. (1966) *Notes on the Analysis of Geographical Distributions.* University of Michigan, Department of Geography, Michigan Inter-University Community of Mathematical Geographers Discussion Paper 8, Part 2, Ann Arbor, Michigan.
10 JENKS, G. F. and BROWN, D. A. (1966) 'Three dimensional map construction'. *Science*, **154**, 857–64.
11 LEARMONTH, A. T. A. and NICHOLS, G. C. (1965) *Maps of Some Standardized Mortality Ratios for Australia 1959–1963.* Department of Geography, Australian National University, Canberra.
12 ARMSTRONG, R. W. (1969) 'Standardized class intervals and rate computation in statistical maps of mortality'. *Ann. Assoc. Am. Geogr.*, **59**, 382–90.
13 ARMSTRONG, R. W. and STORCK, J. (1970) Medical section, *U.S. National Atlas.* Government Printing Office, Washington, D.C., pp. 251–4.
14 MCGLASHAN, N. D. and BOND, D. H. (1967) 'A method for using computer assistance in mapping geographical synoptic data'. *Die Erde*, **98**, 292–7.
15 HOWE, G. M. (1970) 'Some recent developments in disease mapping'. *Roy. Soc. Hlth J.*, **90**, 16–20.
16 ARMSTRONG, R. W. (1969) *op. cit.*
17 HOPPS, H. C. *et al.* (1968) *op. cit.*

18 HOPPS, H. C. (1969) 'Computerized mapping of disease and environmental data'. *Bulletin*, Special Libraries Association (New York), Geography and Map Division, No. 78, pp. 24–31.

19 DROSNESS, D. L., REED, I. and LUBIN, J. W. (1965) 'The application of computer graphics to patient origin study techniques'. *Publ. Hlth Rep.*, **80,** 33–40.

20 HORWOOD, E. M. *et al.* (1963) *Using Computer Graphics in Community Renewal.* Community Renewal Program Guide No. 1, Housing and Home Finance Agency, U.S. Government Printing Office, Washington, D.C.

21 MORRILL, R. L. and EARICKSON, R. J. (1968) 'Hospital variation and patient travel distances'. *Inquiry*, 11, 1–9.

Part II
Public Health Administration

6 The Distribution of Population and Medical Facilities in Malawi*

N. D. McGlashan

The population census of Malawi of 1966 is likely to be of value to many departments of government in the planning of various services offered to the public. To none is it more important than to the Ministry of Health in planning future services or improving existing ones.

Census information can be used in various ways and in this paper a geographical approach will be employed to draw attention to spatial variations in the medical facilities available in Malawi and to assess hospital work loads for purposes of comparison. All statistics used refer to the period mid-1966.

DEMOGRAPHIC DATA

The population enumerated in mid-1966 has been mapped as a dot distribution map (Fig. 6.1). The method was selected for its virtue of simplicity. Whilst certain approximations are inherent in the method of totalling persons to the group required by a single dot and in placing that dot, the method is one which has been internationally accepted.[1] Only in dense areas of rural population such as that shown around Mlanje and Cholo does the method run into difficulties, and these are less serious than the inaccuracies inherent in the presentation of the medical data described below.

As was already well known, this map drawn to illustrate the census shows the very uneven distribution of Malawi's four million people. Large areas of difficult and mountainous terrain in the north and west of the country are very nearly uninhabited, whilst the south-east has an extremely high density of rural population. With large dot size of 500 persons, only three areas, Zomba, Lilongwe and Blantyre-Limbe, are mapped as 'urban', but this is merely a matter of map construction convenience.

* Reprinted from *Central African Journal of Medicine*, (1968) 14, 249–52.

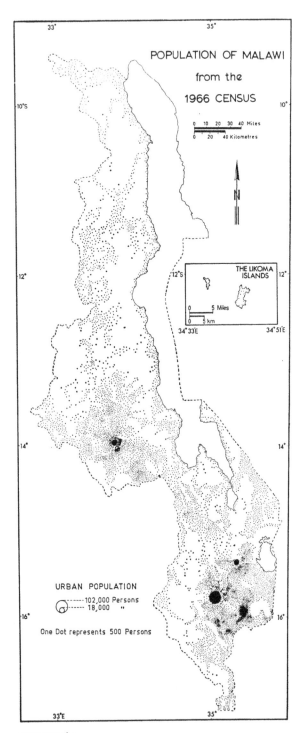

POPULATION OF MALAWI

from the

1966 CENSUS

THE LIKOMA ISLANDS

URBAN POPULATION

102,000 Persons
18,000 "

One Dot represents 500 Persons

FIGURE 6.1

MEDICAL DATA

The measurement of the volume of work carried out by any medical officer or by any hospital is a difficult task. An arbitrary measure which has been used here is the number of general use in-patient beds in any hospital. Doctors in private practice are not considered.

Inaccuracies are inevitable. Amongst these would be hospitals whose staff and facilities are not, in fact, proportional to the number of beds; another would occur if a certain hospital had an out-patient department disproportionately large (or small) when compared to its bed numbers.

The 'sphere of influence' of any hospital, too, is difficult to define. Often the sphere of influence varies with the patient's condition, carcinoma cases, in particular, being frequently referred long distances. In any case, many potential patients within the sphere of influence (however defined) will not present at a given hospital for treatment. These will include those who seek customary remedies or those who, not understanding the nature of their complaint, are put off attending by such practical considerations as the hardship, inconvenience or cost of the journey.

Freedom of choice of the patient is as important in Africa as anywhere else. A patient will be influenced by the reputation (deserved or not) of the doctor and may also 'shop around' for treatment, especially in slow-acting and non-fatal conditions. For example, a patient from Tanzania recently presented at a hospital in the eastern lowveld of the Transvaal for treatment for sterility, a distance of some eighteen hundred miles. Cancer patients too, discharged to their homes as beyond hope, often seek to obtain a more hopeful prognosis in another hospital.

HOSPITAL SPHERES OF INFLUENCE

An attempt to obtain an idea of geographical 'spheres of influence' of hospitals was made in a geographical pathology survey in 1966, which included Zambia, Malawi and southern Katanga.

Resident medical officers were asked from how far in each cardinal direction 90 per cent of their in-patients normally arrived. (The remaining 10 per cent was designed to exclude those who travelled anomalous distances.) Often easily marked political or geographical features were named as the boundary and the influence of transport routes was recognisably important.

In Fig. 6.2 the sometimes conflicting claims of neighbouring hospitals have been adjusted and mapped to show each hospital's normal sphere of influence. The dots from Fig. 6.1 within each area so defined give a figure for total population for comparison with the in-patient facilities (see Table 6.1).

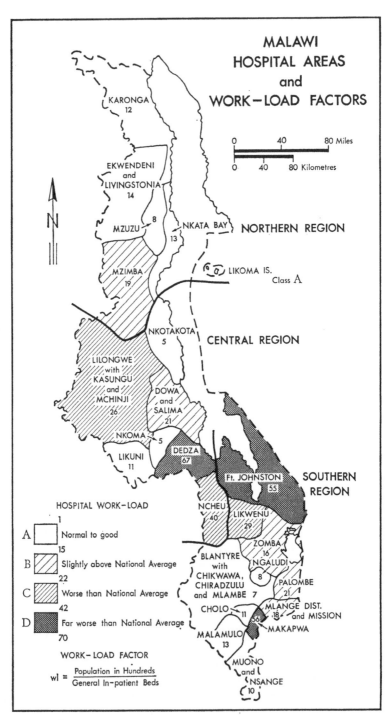

MALAWI
HOSPITAL AREAS
and
WORK-LOAD FACTORS

KARONGA
12

0   40   80 Miles

0   40   80 Kilometres

EKWENDENI
and
LIVINGSTONIA
14

MZUZU   8   NKATA BAY   NORTHERN REGION
                      13

MZIMBA
19

LIKOMA IS.
Class A

NKOTAKOTA   CENTRAL REGION
5

LILONGWE
with
KASUNGU
and
MCHINJI
26

DOWA
and
SALIMA
21

NKOMA   5

DEDZA
67

LIKUNI
11

Ft. JOHNSTON
55

SOUTHERN
REGION

NCHEU
40

LIKWENU
29

HOSPITAL WORK-LOAD

1

A ☐ Normal to good

15

B ▨ Slightly above National Average

22

C ▨ Worse than National Average

42

D ▨ Far worse than National Average

70

ZOMBA
16

BLANTYRE
with
CHIKWAWA,
CHIRADZULU
and MLAMBE 7

NGALUDI
8

PALOMBE
21

CHOLO   11

MLANGE DIST.
18   and MISSION

56   MAKAPWA

MALAMULO
13

MUONO
and
NSANGE
10

WORK-LOAD FACTOR

$wl = \dfrac{\text{Population in Hundreds}}{\text{General In-patient Beds}}$

FIGURE 6.2

TABLE 6.1 *Malawi Hospitals*

Hospital	(a) General Beds		(b) Population Served	b/a	'wl' Factor	Class*
MALAWI REGIONS						
Northern Region	435		544,000	1,251	13	A
Central	790		1,590,000	1,910	19	B
Southern	1,399		1,990,000	1,422	14	A
MALAWI HOSPITALS						
Blantyre	360 ⎫					
Chikwawa	46 ⎪	495	364,000	735	7	A
Chiradzulu	45 ⎬					
Mlambe	44 ⎭					
Cholo	78		85,000	1,090	11	A
Dedza	33		220,000	6,667	67	D
Dowa	75 ⎫					
Salima	20 ⎭	95	195,000	2,053	21	B
Ekwendeni	70 ⎫					
Livingstonia	17 ⎭	87	118,000	1,356	14	A
Fort Johnston	60		331,000	5,517	55	D
Karonga	105		130,000	1,238	12	A
Likoma Is.	40		8,000	200	2	A
Likuni	137		148,000	1,080	11	A
Likwenu (St. Luke)	36		103,000	2,861	29	C
Lilongwe	150 ⎫					
Kasungu	59 ⎬	238	620,000	2,605	26	C
Mchinji	29 ⎭					
Makapwa	14		79,000	5,643	56	D
Malamulo	90		120,000	1,333	13	A
Mlange district	84 ⎫					
Mission	28 ⎭	112	198,000	1,768	18	B
Muono	81 ⎫					
Nsange	50 ⎭	131	125,000	954	10	A
Mzimba	90		175,000	1,944	19	B
Mzuzu	60		46,000	767	8	A
Ncheu	40		158,000	3,950	40	C
Ngaludi	85		71,000	835	8	A
Nkata Bay	53		67,000	1,264	13	A
Nkoma	127		110,000	866	9	A
Nkotakota	120		58,000	483	5	A
Palombe	98		201,000	2,051	21	B
Zomba	200		313,000	1,565	16	B
MALAWI NATIONAL TOTAL	2,624		4,043,000	1,541	15	National Average

* See Fig. 6.2 key for class intervals

From these two figures (columns *a* and *b* in the table) a calculation has been made to show the number of people theoretically served by each in-patient bed. In the next column the 'work-load' factor' defined as:

$$wl = \frac{\text{Population served in hundreds}}{\text{General use in-patient beds}}$$

is tabulated and, in the last column, classed.

This formula devised for 'work-load' is mathematically insensitive in order to allow misjudgements of spheres of influence boundary to make only small differences. For instance, an increase of Cholo's sphere by 20,000 people would increase the wl factor by two points from 11 to 13. On the other hand, with variations of wl (on the mainland) from 5 to 67, extremely large orders of variation clearly occur and can be recognised geographically.

The 'work-loads' are classed by scatter diagram and are based on comparison with the Malawi national average, which is obtained from the fact that some 2,600 beds are available to some four million citizens. Three hospitals in particular seem to provide inadequate facilities in relation to population, and it is worth recording that two of these are small government district hospitals, but also that both serve areas including much difficult and mountainous terrain. Even if the staff and facilities at Dedza and Fort Johnston were immediately *doubled*, their work-load factors would still be in the group 'worse than the national average' (see Fig. 6.2 key). Both of these two cases are aggravated by being geographically contiguous with another hospital (Ncheu and Likwenu) classed as 'worse than the national average'. In practice this means that medical officers can, so to speak, expect little help from 'next door'.

Focus has been thrown upon the worse served areas, for it is here that it would seem prudent for government effort to be concentrated. Whether proliferation of hospitals or increased size of existing ones is a preferable policy is not solely a geographical question. There might seem to be a case for a new hospital to serve the large and remote population to the north-east of Fort Johnston in the Namwera or Mterera area, if this were in accordance with the politico-administrative plans of the country.

It is of interest to note that the Northern Region of Malawi, which is often thought of as remote and underdeveloped in comparison with the rest of the country, compares favourably, for hospital work-loads, with the Southern Region and, especially, is clearly better served on the criteria utilised than the Central Region. The fact that the North has no patient referral or specialist services is an important factor ignored by this method because of its concentration on district level medical services.

The geographical pathology study also shows that, by identical formulae, Zambia's national average 'wl' is six and her worst area, the northern part of Luapula Province, has a wl of twenty-one, which, though bad, is better than two whole groups of Malawi hospitals. A less realistic comparison can be made with England, which is currently aiming to reach an average wl of four.

In a tropical and developing country this work load concept has two weaknesses. Firstly, it fails to allow for the number of out-patient clinics (now about a hundred throughout the land) which effectively reduce the actual load on hospital facilities. Secondly, the types of disease mainly experienced in Malawi, malnutrition, tuberculosis and parasitic infections, are all quite adequately treated by clinic rather than hospital attendance.

This study illustrates the sort of medical purpose that census information in a developing country may properly serve. The method of comparing hospitals by 'work-load factor' may well be improved, but some such means of suggesting areas of population under-served by medical facilities is a necessary study.

ACKNOWLEDGEMENTS

I wish to thank all the medical officers of Malawi who contributed the data which made the survey possible and Miss C. A. Goodbody, B.A., who compiled Table 6.1 and the population map.

REFERENCES

1 WILLIAM-OLSSON, W. (1963) Report of the IGU Commission on a world population map. *Geografiska Annaler*, **45**, 243.
2 Malawi Population Census, 1966, Provisional Report.

7 Flying Doctor Services in Zambia

Mary E. Jackman

INTRODUCTION

The classic study of medical services by geographical techniques was that by Godlund in Sweden in 1961.[1] Even a cursory examination of the data Godlund was able to command shows his method to be clearly unsuitable for African conditions. Nonetheless the techniques he employed may be used as a yardstick and a guide.

Population data in Sweden were so detailed as to provide criteria upon which to forecast population trends fifteen years ahead by four categories from 'decrease' to 'very marked increase'. Only after the September 1969 census has been fully analysed and compared with 1963 figures will Zambia have any such basic population data.

Godlund was also able to map hospital 'spheres of influence', not so much by where patients actually came from, but by regionalisation based on travel time and cost, to give a rationale to explain patients' journeys to medical services.

In Zambia, hospital spheres of influence have been described for the situation in 1966.[2] In a country where public transport is (for rural areas) almost unknown and, even in the capital itself, only 10 per cent of patients arrive at clinics by motor transport,[3] it may be claimed that *actual* patient journeys provide a more realistic criterion than measures of speed or cost potential along different classes of road.

In the study to be described here the prescription was simple – as a requirement, if not in solution. *The Flying Doctor Service should so site its clinics as to ensure that every person in Zambia lives within thirty-five miles of medical services with a doctor.* This distance represents 'government policy' and is said to be an average day's bicycle ride – presumably by a healthy adult. It is, however, a distance that has proved convenient in the provision of other rural development services.

THE POPULATION OF ZAMBIA

Zambia covers 290,586 square miles, with the relatively small population of about four million people in 1968. As in many African countries the great

majority of the population (87 per cent) are rural people. The country therefore poses serious problems in the provision of medical services. Not only
is the population unevenly distributed throughout the area, but the internal
movements of people are not yet fully known.

In 1963 the first full census of the African population of Zambia (Northern Rhodesia)[4] was taken. Kay[5] has provided a critique upon it. A further
census, taken in 1969, will allow a more complete study of population movement than has hitherto been possible. It is already evident that the movement of peoples into urban areas is greatly in excess of the estimated rates,
but no overall study of the movement of rural population is yet available.
The 1963 census revealed a total population of under four million, of whom
800,000 lived in the major urban areas. These areas lie mainly on the
Copperbelt and along the line of rail. Of the remaining rural population
Kay identified three qualitative density categories which he described as
'densely populated', 'sparsely populated', and 'virtually uninhabited'.

The areas which fall within the 'densely populated' category comprise
the Upper Zambesi Valley (Mongu, Kalabo, Balovale), Southern Province
(Choma, Gwembe, Mazabuka), Eastern Province (Petauke, Chipata, and
part of Lundazi), Luapula Province (Kawambwa, Mansa, Samfya),
Northern Plateau (Mbala, Kasama), and a central area (Lusaka, Mumbwa).
These areas together hold 47 per cent of Zambia's population on only 12
per cent of its area. At the other extreme of density the country's two
major 'virtually uninhabited' areas centre on the two largest game parks;
the Kafue National Park and the Luangwa Valley Park. In these areas,
16 per cent of Zambia's area holds only 10,000 people, 0·25 per cent of the
whole population.

Between these two extremes the population is spread at an average
density of seven persons per square mile. There are, of course, many variations in the density and types of distribution within these areas. The actual
distribution is affected by many factors, some purely physical such as the
need for a perennial water supply, and some involving history and culture.
The more sparsely populated areas, for instance, still seem to represent the
remnants of the 'no-man's land' between major tribal groupings. Com-
munications also pose some problems, for although the situation is improving rapidly, the internal transport network is still far from adequate,
and distances involved are considerable.

MEDICAL FACILITIES

The existing pattern of medical services consists of several different types
of provision.[6,7] Firstly there are the hospitals with resident medical officers.
These are found mainly, but not entirely, in urban areas. Secondly there
are the hospitals with a hospital assistant or nursing sister in charge

which are visited regularly by a nearby doctor. Both these types of provision may be run by government, industry, or mission services. Thirdly there are rural health services. These are clinics, dispensaries and health centres situated in rural areas, and are visited regularly by medical staff. Apart from these 'ground-based' services, there are the five separate airborne doctor services. These serve different areas and are run by government, mission and charitable organisations. They usually consist of an airstrip with a clinic site nearby which is visited by a flying doctor 'plane.

A METHOD FOR ESTABLISHING FLYING DOCTOR SITES

From the beginning it proved impossible to consider the work of the Zambia Flying Doctor Service (Z.F.D.S.) except within its context as part of the total medical service. Equally it proved impossible to recommend sites for expansion without reference to the expansion which may occur within other branches of the national medical services. The subject was therefore treated by viewing the medical service as a whole and by giving special reference to the place of Z.F.D.S. within it.

As the problem was one of spatial distribution of medical services, and the criterion was that every member of the population should be brought within a fixed distance of a doctor's services, the first step was to exclude the areas already served at this level. This was done by marking, on a population dot map of Zambia, the positions of the hospitals with doctors and existing flying doctor stations. These were then credited with a thirty-five mile radius area-of-service (Fig. 7.1). In this way it was possible to distinguish areas already well or badly served. Perhaps the most noticeable result of this exercise was to show up the lack of services in the highly populated areas of northern Zambia, which tended to justify the Z.F.D.S.'s present heavy commitment there. Areas of western Zambia are comparatively well served and this may be explained by the fact that many missions historically concentrated on these remote and inaccessible parts. Fig. 7.1 shows that many of the doctor services in this area too are flying doctor services. When all the areas already served are excluded, the resulting 'un-served' population is only about 900,000, less than a quarter of Zambia's total population. This evidences a generally excellent spread of services.

In recommending future sites for expansion of medical services it was necessary to attempt to cover the unserved regions economically and with the least possible duplication. For this reason the idea of two lake doctor services was again put forward.[8] It was suggested that a fully equipped launch should be based on a lakeside hospital to visit several new clinics along each lake's shore. This was recommended as an effective and economical way of serving the dense populations, estimated to total about 290,000 people, living within thirty-five miles of the shores of Lakes Mweru and

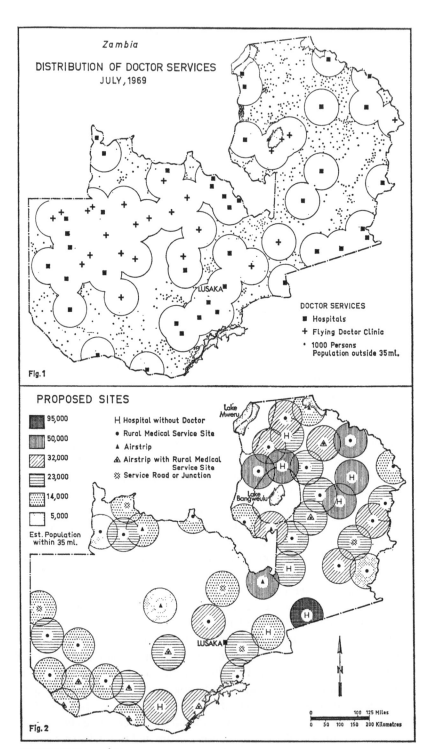

Fig. 1

Zambia

DISTRIBUTION OF DOCTOR SERVICES
JULY, 1969

DOCTOR SERVICES

■ Hospitals

+ Flying Doctor Clinic

· 1000 Persons
Population outside 35 ml.

LUSAKA

Fig. 2

PROPOSED SITES

95,000

50,000

32,000

23,000

14,000

5,000

Est. Population
within 35 ml.

H Hospital without Doctor

● Rural Medical Service Site

▲ Airstrip

⊿ Airstrip with Rural Medical
Service Site

✳ Service Road or Junction

Lake Mweru

Lake Bangweulu

LUSAKA

N

0   100 125 Miles

0   50 100 150 200 Kilometres

FIGURES 7.1 and 7.2

Bangweulu. At this point, therefore, all parts within this distance of the lake shores were divided from the rest of the map, as being areas for which recommendations could straight away be made.

The remaining medically unserved areas were then found to fall into ten major divisions. For these divisions it was necessary to recommend either sites for future flying doctor clinics or sites for expansion of ground-based services. Several methods of procedure were entertained but, in order to eliminate human error and to be objectively certain that the number of sites with which the area could be covered was, in absolute terms, the most economical, it was decided to use a computer to select those needed for development. To achieve this, the coordinates of each unserved population dot from Fig. 7.1 were fed into an IBM 190 computer and this was then programmed to recommend the smallest number of new sites needed to cover all the population at a distance of up to thirty-five miles, and to print out the coordinates of the population dots which could be used as sites. This method, as well as ensuring the least possible overlap of services, meant that each computer-recommended site was an approximate centre of gravity of population.

The sites chosen by computer were then considered subjectively in terms of their practicalities and sites were re-adjusted where necessary taking five considerations into account:

(1) The adjustment should not prejudice the overall network.
(2) Maximum population should live immediately around the clinic in order to reduce patients' travelling time.
(3) Existing airstrips should be used wherever possible.
(4) Existing health service facilities should be used wherever possible.
(5) New airstrip sites should be on or near to existing surface routes to facilitate the supply of heavy stores.

The results of this are summarised as Fig. 7.2. Although considerable overlap appears between some pairs of sites, no site can be eliminated without leaving gaps in the pattern of service. In fact most overlap occurs north of Bangweulu where the population density warrants a closer network of facilities. The population which would be served by each recommended site was also computed to give a guide to an order in which the recommendations might be implemented. It will be observed that these estimates of population served at each site are considerably larger than the total of dots from Fig. 7.1. This is because each site has had attributed to it, not only the currently unserved rural population out to a radius of thirty-five miles, but also population within this distance which is at present served by existing ground hospitals. That is, in all cases of overlap of service radii, the doubly served population is counted as tributary to *both* sites. Any other procedure would invalidate the method of putting sites into an order of priority. The heavier categories of shading on Fig. 7.2 thus indicate areas of most urgency on the basis of greatest potential population to be served.

The recommended sites can be divided into five distinct types:

(1) Sites with existing hospital facilities but no doctor.
(2) Sites with existing rural health service facilities but without a doctor.
(3) Sites with existing airstrips.
(4) Sites with both airstrip and rural health service facilities.
(5) Sites with only road junction or road access to recommend them.

RECOMMENDATIONS

From these categories, it was possible to recommend an ordered expansion of flying doctor facilities (in conjunction with other branches of the national medical services) into a pattern of doctor services which would cover the whole of Zambia and bring every member of the population within thirty-five miles of a doctor.

OTHER APPLICATIONS

This type of approach, described here in relation to the provision of medical services, is one that might be used for many other problems where there is a need to bring some commodity or service within the reach of a scattered population. Other fields where it might prove useful are the provision of agricultural services, tractor services and maize collection points, or the provision of commercial services such as banking. If a number of 'services' were to coincide at certain sites these would gradually provide natural foci for rural population re-grouping.

REFERENCES

1 GODLUND, SVEN (1961) *Population, Regional Hospitals, Transport Facilities and Regions in Sweden*. Lund Studies Series B, No. 21.
2 MCGLASHAN, N. D. (1968) 'The distribution of population and medical facilities in Zambia'. *Med. J. Zambia*, **2**, 17–25.
3 MCGLASHAN, N. D. (1968) 'The geography of Lusaka City clinics'. *Med. J. Zambia*, **2**, 147–53.
4 Government Printer, Lusaka (1964) *Census of African Population of Northern Rhodesia*.
5 KAY, G. (1967) *Maps of the Distribution and Density of African Population in Zambia*. University of Zambia, Institute for Social Research Communication No. 2.
6 Ministry of Lands and Mines, Lusaka (1967) *Hospitals in the Republic of Zambia* (Map) and (1969) *Medical Facilities in Zambia*. National Sheet Atlas, Map No. 14.

7 MCGLASHAN, N. D. (1967) *The Distribution of Six Diseases in Zambia.* National Sheet Atlas, Map No. 26.
8 MCGLASHAN, N. D. (1965) *The Distribution of Blindness in the Luapula Province of Northern Rhodesia.* Unpublished University of London M.A. thesis. Presented to Zambia Ministry of Cooperatives, Youth and Social Services in April, 1965.

8 Problems of Public Health among Pastoralists: A Case Study from Africa*

R. M. Prothero

INTRODUCTION

Traditionally nomadic pastoral peoples have been resistant to change. Reserve and a seemingly inate conservatism have inhibited the processes of modernisation in whatever forms these may have appeared. The majority of nomadic pastoralists maintain themselves precariously in the environments in which they are located, with ways that have remained little changed over centuries. To a considerable extent the nature of these environments has limited the amount of change which could take place. But even where official policies for settlement and improvement have been applied they have been met in some instances with resistance and a degree of coercion has been necessary to make measures effective.[1]

In response and in adaptation to harsh environments, with a continual if not continuous need to seek for pasture and water, the element of mobility, which is dominant in the lives of nomadic pastoralists, is unfavourable towards the acquisition of amenities and facilities which are associated with more settled ways of life. There are major problems in the provision of health and education services which are normally designed for populations whose patterns of activity are relatively much more rigid and confined in terms of space and time.

Evidence of the state of health of nomadic pastoralists is scanty. Overall medical services in Africa are limited, pastoralists are very often minority groups in the total population, and, because of difficulties of making contact with them, inevitably they have tended to receive only very little, if any, of the services available. The methods of indexing in the literature of medical abstracts are such as to make it difficult to identify those studies which have been undertaken. Social anthropologists, who have probably paid more attention to nomadic pastoralists than workers in any other

* This essay is based to a considerable extent on 'Public health, pastoralism and politics in the Horn of Africa', the Sixth Melville J. Herskovits Memorial Lecture which was delivered under the auspices of the Program of African Studies, Northwestern University, U.S.A., in 1967. The lecture was published in 1968 under the same title by the Northwestern University Press, Evanston, to whom due acknowledgement is made.

discipline during the last twenty-five years, have given little of their attention to health matters. It is therefore possible to quote only generalised
statements, which are of limited validity only. Furthermore, these are
sometimes contradictory.

For a group of Fulani nomadic pastoralists in north-eastern Nigeria,
Stenning noted that they were exposed to a rigorous life in a harsh environment in a variety of ways.[2] Accident and injury (snake bite, for example)
were common. Exposure, to wetness and to relative cold in poorly constructed dwellings, produced muscular and respiratory complaints, and
could be identified as a major factor in high infant mortality. Filarial and
helminthic diseases from contact with polluted water were prevalent. However, diet was more varied and nutritionally more balanced than that of
sedentary grain cultivators in the same area, and Stenning commented on
the marked absence among nomadic children of skin diseases, sores and
ulcers compared with the children in sedentary villages. Overall he noted
among this Fulani group a general lack of sympathy for the sick and con-
genitally deformed who are unable to cope, observing that 'the demands
of the pastoral life are in general inimical of quick recovery from more
serious illness'.

Among nomads in the Sahara desert malaria and syphilis are reckoned
to be no longer the scourges which they were previously, but gonorrhoea
remains common. Eye diseases, particularly trachoma, often associated with
nomadic groups are deemed to be more prevalent among sedentary oasis
communities. Opinions have varied as to the prevalence of tuberculosis,
assessing it variously as a serious problem[3] and as infrequent.[4] This disease
has been noted as common among nomadic Bedouin tribes in southern
Israel[5] and among nomads in Somalia[6]. However, the relative ease and
efficiency of BCG vaccinations available during the last decade may well
have ameliorated conditions in respect of this disease.

Data on fertility among nomadic populations in Senegal,[7] Mauretania[8]
and Sudan[9] point to rates which are considerably lower than those among
sedentary populations, and in the case of the Sudan the trend of rising
fertility among the latter is absent in the former. In identifying the causes
of these differences Henin specifies differential marriage patterns (later
marriage, higher marital instability, and higher rates of polygamy among
nomads) and medical/psychological factors.[10] For the latter he notes that
'data on endemic diseases in the Sudan in general and for nomads in particular is scanty', but suggests important factors are (1) a high incidence
of venereal disease (gonorrhoea and syphilis) caused in part by extra-marital
relationships, particularly through separation of wife and husband during
certain periods of the year; (2) birth prevention practices, especially abor-
tion; and (3) miscarriage encouraged by malaria infection and the absence
of medical services. These are accentuated by the strenuous life which

nomadic women lead and associated low standards of hygiene. In contrast to Stenning, Henin comments on the lower nutritional standards of the nomadic population in the Sudan as compared with the settled population.

The case study which follows is on a macro-scale in all respects. It is concerned with the problems of public health among nomadic pastoralists in the Horn of Africa, the only large area of the continent where they form a majority of the population – between 75 and 85 per cent in the Northern Regions of Somalia (former British Somaliland) and between 60 and 65 per cent in the Southern Regions (formerly the U.N. Trusteeship Territory of Somalia). Attention is directed particularly to malaria, the major disease responsible for high morbidity here as elsewhere over much of Africa. Otherwise the range of diseases found in the Horn is probably more limited than many other parts of Africa, undoubtedly linked with the prevailing aridity.

ENVIRONMENT AND DISEASE IN THE HORN OF AFRICA

The north-eastern extremity of the continent includes parts of the Ethiopian Federation, the Territory of the Afar and Issa People (formerly French Somaliland) and north-eastern Kenya, and the whole of the Republic of Somalia. From similarities in the physical environment and in ethnic distribution there is justification for considering these together, though this is not to imply a uniformity of either physical or human conditions over the whole of the Horn, for this is far from being the case.[11] The region is bounded on the west by the edges of the high Ethiopian plateau, a series of steep, rugged escarpments deeply incised with river valleys, rising to over 2,000 metres above sea-level from the lower-lying plateaus and extensive lowlands which extend to the Gulf of Aden and to the Indian Ocean (Fig. 8.1). In places these lowlands give way to areas of highland, which in the Northern Regions of Somalia rise to more than 2,000 metres above sea-level. The Horn, both lowland and highland, is characterised by relative and absolute aridity, with annual rainfall everywhere averaging less than 750 mms. In the basins of the Webe Shebeli and Juba rivers which rise in the Ethiopian plateau the availability of water is relatively better than elsewhere in the Horn. All these and other features of the physical environment influence the occurrence of disease both in space and in time.

Malaria is influenced in its incidence and intensity by physical conditions. It occurs commonly up to 2,000 metres, but above this altitude on the Ethiopian plateau temperatures generally inhibit the breeding of the mosquito vector.[12] At lower altitudes, in the river valleys particularly, malaria is more prevalent, with a transmission season extending from July to December and the peaks of infection occurring in October and December. The major vector, *A. gambiae*, has shown itself to be particularly tough and adaptable, so that the few adult vectors which survive extremely arid conditions breed abundantly when sufficient rain falls to extend the breeding places. With a reservoir of infection and a highly susceptible human population, epidemics occur.[13]

Relatively minor modifications in the environment through human action have extended the conditions in which the vector may breed. These have occurred on the Haud plateau particularly, an area of largely featureless

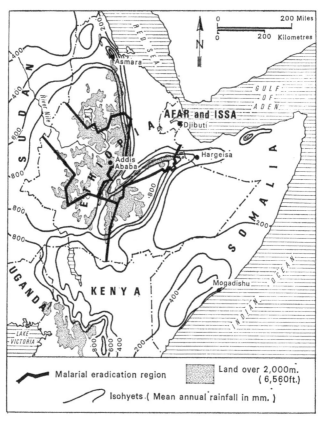

FIGURE 8.1 Relief and rainfall in the Horn of Africa with malaria eradication regions in Ethiopia.

rolling surfaces ranging from 600 to 1,500 metres above sea-level and extending over 125,000 square kms. in both Somalia and Ethiopia. For about eight months of the year the Haud is naturally waterless, without permanent watercourses and with underlying strata which have not yet yielded water even at great depth.[14] Water collects in shallow depressions (balleh) on the surface during the main (Gu) rains in April/May and to a lesser extent during the secondary (Dhair) rains in September/October. It is rapidly lost through percolation and evaporation and by human and animal consumption, so that in the past the Haud was habitable only for limited periods of the year. Improvements in water availability have been effected by enlarging and deepening natural depressions to form small reservoirs, but more particularly by the construction of 'tanks' to collect water during the rains and to store it for use during the dry season (Fig. 8.2). The time during which the Haud is habitable has thus been extended.

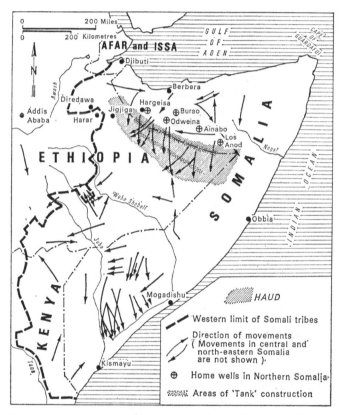

FIGURE 8.2 Movements of Somali pastoralists in the Horn of Africa.

In this way the problem of water shortage has been alleviated, but a new problem has been created by extending in space and in time the environments in which malaria vectors may breed. Attention has been drawn to the contrast between the situation in the Mudugh and Mijerteynia Provinces of Somalia, where the water in deep wells provides only scattered and scanty foci of vector breeding, and the much wider spread of 'tanks' in the Haud.[15] The *A. gambiae* has spread to the latter from its natural breeding places in the Webe Shebeli basin. The water surfaces in the 'tanks' may be treated with a film of oil to prevent mosquito-breeding and this treatment began in what was the British Protectorate of Somaliland in 1954. However, the 'tanks' are located also in the larger section of the Haud which is in Ethiopia and have been constructed up to distances of 80 kms. and more from the boundary with northern Somalia. Certainly during the early years of the 1960's complementary treatment of 'tanks' in the Ethiopian section of the Haud was not being undertaken and mosquito-breeding was able to continue in them.

THE HUMAN ELEMENT: PASTORAL MOBILITY

Thus far in outlining the malaria situation in the Horn of Africa there has been only passing reference to people. With parasites and vectors they are the vital elements in human malaria. They may be identified as both directly and indirectly the most difficult element with which to deal. Of major concern here is their characteristic of mobility, but many other of their activities are also relevant.

The basic distinction between the high Ethiopian plateau surfaces and the lower-lying lands to the north and east is reflected in basic contrasts in ways of life. The people inhabiting the well-watered plateau are primarily sedentary cultivators, while those at the lower and more arid altitudes are primarily pastoralists.[16] There are of course exceptions to this generalisation, as with the sedentary or semi-sedentary cultivators in the Borama region of north-west Somalia and in the Webe Shebeli and Juba valleys and other better-watered parts of southern Somalia. Basic contrasts in ways of life are complemented by ethnic and cultural differences. The limitations which aridity imposes upon economic activity are considerable, and nomadic pastoralism, with a continual search for pasture and water, is the only means of livelihood possible for the majority of the population. The people who follow this way of life are of the Samaale section of the Somali, a relatively homogeneous ethnic group linked by a common language, adherence to Islam, and a social system with a common heritage and with kinship relationship strongly emphasised.[17] Their movements facilitate the transmission of malaria and present major problems in programmes for eradication of the disease.

The primary loyalties of Somali pastoralists are not naturally to land, to which they have little direct attachment and with which their relationships are impermanent and continually changing in response to the availability of pasture and water. The nature of the physical environment is such that the whereabouts of people with their stock cannot be predicted with precision in either space or time. In the vast areas of the Horn of Africa over which pastoralists are distributed, at very low average population density, it is possible to demarcate only the approximate extent of the grazing grounds of the various Somali tribes. The available evidence indicates a high degree of variation from year to year and from season to season within each year.[18]

Though Somali ties are not with locality, paradoxically it is with the spatial aspects of their life as expressed in long-term, annual and seasonal movements that there is need for concern. Over the course of several centuries Somali pastoralists with their animals have dispersed over an increasingly wide area of the Horn of Africa.[19] Increasing numbers of people and stock with increasing demands for food and water have almost certainly been important factors which have influenced this dispersion. The present distribution of Somali extends westward to the bounding escarpment of the Ethiopian plateau and southward into north-eastern Kenya (Fig. 8.2).

Within this area and in which they form the majority of the population, Somali pastoralists move annually and seasonally in search of water and pasture for their stock. To some extent these movements may be described as cyclic; they conform to a general pattern which is repeated each year. In southern Somalia pastoralists are concentrated with their animals in the basins of the Webe Shebeli and Juba rivers in the dry periods of the year and are more widely dispersed during the wet season. In northern Somalia in the dry seasons they concentrate near deep wells providing permanent water, as in the vicinity of Hargeisa, Burao, Odweina, Ainabo and Los Anod (Fig. 8.2). But when the main *Gu* rains, and to a lesser extent the secondary *Dhair* rains, fall in the spring and autumn respectively, the pastoralists are able to move southward into the Haud. From it they must subsequently retire, particularly during the long dry season, *Jilaal*, which lasts from November to March.

Within the basic annual cyclic patterns of movement there are more limited (in space but not in number) and more complex seasonal movements of people and of stock. These occur in the Haud in the wet season in response to the detailed and immediate availability of water and pasture which are determined by the sporadic occurrence of rain in storms of limited duration and affecting only limited areas, of perhaps a few square kilometres. The different groups compete with one another for these scarce commodities. To obtain them they are capable of rapid mobility, for example covering distances of more than 160 kms. in sixty hours. To increase mobility, groups divide, so that the young men with the camels, which form the major stock element, may range over much greater distances than older people with the children who tend the sheep and goats. But the mobility of the latter groups is only of a relatively lesser degree.

In addition to these movements, which are conditioned essentially by environmental factors, there is a further element of mobility which is social in nature but which has distinct spatial manifestations. The basic social unit among the Somali, *rer*, consisting of three or four families, is in a state of constant flux; there is continual fission and fusion as families break off from one *rer* to join another. These changes in basic social groupings are related to the nature of extended kinship ties.

In many respects, therefore, the Somali nomadic pastoralists are highly mobile, as compared with neighbouring sedentary cultivators, and their mobility is greater than that found among the majority of people in Africa. This mobility raises major problems for general social and economic developments, which are more restricted in the pastoral areas of the Horn of Africa than in most other parts of the continent. Mobility is a particular problem for public health.

PASTORALISM AND PUBLIC HEALTH

The mobility of the population is a factor in the transmission of malaria through the movement of persons infected with parasites providing a reservoir for the infection of others. There is also some evidence to suggest that adult anopheles mosquitoes may be transported in the belongings of nomads as they move from one grazing area to another. The latter may explain the spread of *A. gambiae* northward from the Webe Shebeli basin. In an area of considerable aridity like the Horn of Africa the problems of man/mosquito contacts are somewhat similar to man/tsetse fly contacts in the transmission of trypanosomiasis. Both host and vector are seeking the limited supplies of water that are available and will therefore inevitably become juxtaposed one with another.

As a factor in the total malaria situation, pastoral mobility is more serious in its limiting effects upon measures for control and eradication of the disease. Some success against malaria was claimed in the 1950's in Mudugh and Mijerteynia Provinces of the then U.N. Trusteeship Territory of Somalia, through the use of residual insecticide against the vector coupled with the administration of antimalarial drugs to the population.[20] This progress was made in conditions which have already been noted as relatively more favourable than those in the Haud. In the Somaliland Haud, residual insecticide spraying, coupled with drug distribution, was carried out in the last half of the 1950's and into the early 1960's without any conspicuous success.[21] Mobility and intermixing of the population made it impossible for insecticide spray teams to make effective contact with people and to obtain total coverage of dwellings. To this problem must be added the greatly increased costs, in time and energy of personnel, and of transport and materials, with the distances that have to be covered to locate settlement which is not just dispersed but also mobile and impermanent in terrain where even the few existing roads are difficult to negotiate at all times.

The transportable nomadic dwelling (*akal*), roughly hemispherical in shape, constructed of a framework of branches and covered with a combination of mats, skins, cloth, and a variety of other materials, is largely unsuitable for spraying with insecticide, even assuming that it can be located. The materials of which it may be made often do not retain the residual insecticide even under static conditions. When frequently transported, in the process of striking, moving, and re-establishing camp, the losses are even greater. Added to this is the fact that surfaces which formed the interior of an *akal* at one camp and were sprayed with insecticide may become the exterior surfaces at the next site and the insecticide may be washed away in one downpour of rain.

Success with chemotherapy and chemoprophylaxis requires the maintenance of an adequate supply of drugs, the efficacy of which is limited in time and which must therefore be distributed and taken regularly. It is a major problem to ensure such regularity among populations which are even relatively sedentary. Where only a minority of the population live in permanent dwellings in permanent settlements, permanent public health centres (hospitals, clinics and dispensaries) are likely to be of service to very few people; many for the greater part of their lives may literally never come near them. Mobile dispensaries have been used which can be set up where relatively dense concentrations of nomadic settlement develop temporarily; and then can be moved elsewhere when these disperse.[21]

Some success was achieved also in British Somaliland with the establishment of a network of tribal aides to provide liaison between their people and the medical authorities for the treatment of malaria and for other aspects of public health. There is undoubted need for the further development of such an organisation.[22] One of the difficulties that has arisen here is related to the highly democratic nature of Somali society, which has been described as 'a pastoral democracy'.[23] Individual rights are of paramount importance and under these circumstances it was difficult to find anyone who could be regarded as representative of a group. With social and political developments under an independent government in Somalia since 1960 these circumstances may be changing. The Somali have shown themselves to be remarkably receptive to modern political influences, and there may be a more rapid erosion of traditional social attitudes than might have been expected. But it is unlikely to be a very rapid process and the problems of developing and maintaining adequate contact and cooperation between the population and the public health authorities will remain for some time, even though probably with diminishing importance.

POLITICS AND PASTORALISM

The problems for public health created by pastoralism alone would be of major importance, with no easy solutions offered. But these problems are much increased by political factors, particularly in respect of the relations between the Republic of Somalia and its neighbours. These relations derive from the political circumstances not only of the present but also of the past – both the recent past (the late nineteenth century and the present century) and the more distant past. They are conditioned in part by the nature of both long-term and short-term pastoral movements. They are related to the colonial situation that formerly existed in this part of Africa and to its aftermath. The political situation and its various implications are examined here only in the broadest outline. Its many facets have been examined in more detail in a number of studies during the past decade.[24]

The long-term migratory drift of Somali pastoralists over many centuries brought them into contact and sometimes into conflict with the sedentary cultivators occupying highland areas in the western parts of the Horn. Somali groups were still pressing southward in the early decades of the present century. Upon this dynamic distribution international boundaries in the Horn of Africa were superimposed, dividing the Somali people among the colonial administrations of the French, British and Italian Somalilands and Kenya, and the Empire of Ethiopia. In some instances the boundaries between these territories were agreed on and defined with relative clarity; in others they were not agreed on and to the present day have remained vague and indefinite.[25] Between Ethiopia and Somali territory, over which British protection had been established in the 1880's, the agreed boundary did not accord with the ethnic distribution of Somali people, but even more seriously it divided the Haud plateau so that the major portion of this region fell in Ethiopian territory.[26] Provision was made for the tribes to move seasonally into Ethiopian territory to graze their stock, at which time they would be subject to Ethiopian jurisdiction.

Demands for the unification of Somali people culminated in 1960 with the independence of British Somaliland and of the U.N. Trusteeship Territory of Somalia and their immediate unification in the independent Republic of Somalia. The existence of a Somali state which includes a majority of Somali people has provided a focus for aspirations to incorporate all Somali and the areas which they occupy in the adjacent territories of French Somaliland, Ethiopia and Kenya. The opposition of each of these adjacent territories to such a nationalist aim has resulted during the last decade in tension and various forms of conflict between Somalia and her neighbours.

Some very fine points of international law are involved. Problems relate to (1) the delineation of boundaries in the colonial era without adequate knowledge, and with little or no reference to the peoples immediately concerned; (2) the precise rights of the colonial power at the time when boundaries were agreed and the rights of successor states; and (3) the difficulties of rigid boundaries with peoples who do not recognise rigid space relationships.

POLITICS AND PUBLIC HEALTH

This political background of dissatisfaction, distrust and open conflict has had a fundamental impact upon the public health problems of the Horn of Africa and in every respect militates against their solution. Here, as elsewhere in Africa, disease recognises no political boundaries. Interterritorial cooperation and coordination are essential in measures to improve public health. Successful malaria eradication, for example, can be conceived of realistically only in terms of at least the whole of the Horn of Africa.

The likelihood of any cooperation and coordination being achieved at the present time is minimal. Reference has been made previously to the absence of treatment to prevent vector-breeding in 'tanks' in the Ethiopian section of the Haud at a time when treatment was being effected in the adjacent areas of Somaliland. Mosquitoes, even less than people, do not take note of boundaries. Movements of population at the scale on which they occur in the Horn of Africa, particularly during the seasons of malaria transmission, lead to the wide dispersal of infected persons and thus increase the spread of infection where there are vectors present. The difficulties of contact with people for treatment have been considered; they are much increased if, for periods of the year, large sections of the population may be in another country where no antimalarial measures are being taken. Under such circumstances the good work that may be achieved under difficulty in one country may be completely undone by the absence of at least comparable, though preferably coordinated, action in those adjacent to it. Africa and other malarious areas of the world provide too many depressing examples of the absence of such cooperation and coordination.[27]

Circumstances in the Horn of Africa are not unique, but in many respects they offer extreme examples. Before independence for Somalia there was no coordinated action between the medical authorities in British Somaliland and in Ethiopia in antimalaria operations, and even W.H.O. personnel working in the former were not allowed to pursue their investigations across the boundary into the latter. Malaria work through the W.H.O. was further impeded for many years by an administrative anomaly; Ethiopia and the U.N. Trusteeship Territory of Somalia formed part of the W.H.O. Eastern Mediterranean Region while British Somaliland, due to its colonial connections, was part of the W.H.O. African Region. This has now been

E

corrected and the whole of the Republic of Somalia is in the W.H.O. Eastern Mediterranean Region.

Recognising the immense problems involved when factors such as those in the Horn of Africa have to be taken into account, the strategy for malaria eradication has changed there and elsewhere in Africa. Ethiopia and Somalia are now both engaged in the pre-eradication phase in their respective programmes of malaria eradication. This should entail a full evaluation of the elements and factors in the malaria situation – parasite, vector and human host (though it is unlikely that sufficient attention is being devoted to the last of these) – to enable a rational programme of eradication to be designed. In this phase also an infrastructure of trained personnel and equipment is built up to operate the programme in its successive stages. But the pre-eradication surveys and the programmes for malaria eradication in Ethiopia and Somalia have been designed independently of one another. As in most other African countries the magnitude of the task in terms of the resources required is such that Ethiopia can only consider a phased programme of operations, dealing with one region of the country at a time[28] (Fig. 8.1). It is questionable if such an approach will succeed where population mobility is characteristic, for the evidence suggests that a simultaneous operation on a wider scale is required. However, since there is no alternative to adopting this phased approach it is a matter of regret to find that the order in which the regions of Ethiopia are to be tackled places last that part of the country which is in the Horn of Africa and adjacent to the Republic of Somalia. In view of the political circumstances outlined the decision is not surprising, but in view of the other factors in the malaria situation in the Horn of Africa it may well be disastrous for successful eradication of the disease. Any favourable results which might be achieved in Somalia could be seriously prejudiced, if not nullified, by the absence of simultaneous and coordinated action in adjacent parts of Ethiopia. Political relations between the two countries have improved during recent years, but there is no guarantee that this improvement will be maintained since the fundamental causes of difference between them remain unresolved.

CONCLUSION

In this study it has been possible to do no more than outline some of the problems which pastoralists present in public health, and to demonstrate the ways in which such seemingly disparate elements as public health, pastoralism and politics are intimately linked. The more detailed facets of these problems are even more heterogeneous in character; they include the history of the recent and the more distant past, ethnic distributions, social organisation, population structure and mobility, economy, physical environment, disease epidemiology, vector entomology, medical services, international law and politics. Each of these in turn has more detailed facets, which are linked with one another in complex relationships.

The Horn of Africa may well present an extreme case, but elsewhere in Africa public health problems, involving malaria and other diseases, are not dissimilar in their multidisciplinary character. Success in solving them will not be achieved until this character is more fully recognised and until inter-disciplinary cooperation is established to analyse and understand them more fully than has been the case in the past.[28] The geographer may contribute significantly to this work for the majority of the factors involved have spatial/temporal dimensions. The macro-scale approach which is presented in this study is necessary and the wide-ranging, overall view should be maintained, but in terms of planning for action it needs to be linked with investigation at a more detailed scale. Work of the latter nature has been undertaken to only a very limited extent in Africa.[29]

REFERENCES

1 UNESCO (1959) 'Nomads and nomadism in the Arid Zone'. *Int. Social. Sci. J.*, **11**, 481–585.

2 STENNING, D. J. (1959) *Savannah nomads*. O.U.P. London.

3 Comité de Coordination Scientifique du Sahara (1960) *Journées d'Informations Medico-Sociales Sahariennes*. Paris.

4 DOURY, P. (1959) 'Le Hoggar: étude medicale'. *Archs Inst. Pasteur d'Algérie*, **37**, 104–64.

5 BEN ASSA, B. J. (1960) 'Vital statistics concerning tuberculosis among Bedouin tribes in southern Israel'. *Israel med. J.*, **19**, 69–73.

6 YOUNG, K. (1957) 'Impressions of tuberculosis in Somaliland'. *Tubercle*, **38**, 273–9.

7 U.N. Economic Commission for Africa (1964) *African Seminar on Vital Statistics*. Addis Ababa.

8 U.N. Economic Commission for Africa (1965) *Econ. Bull. Afr.*, **5**, Table B.19.

9 HENIN, R. A. (1968) 'Fertility differentials in the Sudan (with reference to the nomadic and settled populations)'. *Popul. Stud.*, **32**, 145–64.

10 HENIN, R. A. (1969) 'The pattern and causes of fertility differentials in the Sudan (with reference to nomadic and settled populations)'. *Popul. Stud.* **23**, 171–98.

11 PROTHERO, R. M. (1969) 'North-East Africa: a pattern of conflict'. In R. M. Prothero (ed.) *A Geography of Africa: Regional Essays on Fundamental Characteristics, Issues and Problems*. London.

12. CHAND, D. (1966) 'Malaria problem in Ethiopia'. *Ethiopian med. J.* **4**, 17–34.

13 CHOUMARA, R. (1961) 'Notes sur la paludisme au Somaliland'. *Riv. di Malariologia*, **40**, 9–34.

14 STOCK, S. (1959) *Water supply and geology of the Haud Plateau, Somaliland Protectorate*. Unpublished report. Public Works Department, Hargeisa.

15 MAFFI, M. (1960) 'La malaria nelle regioni de Mudugh e delle Migiurtinia, Somalia'. *Riv. di Malariologia,* **39,** 21–118.

16 LEWIS, I. M. (1955) *Peoples of the Horn of Africa: Somali, Afar and Saho.* International African Institute, London.

MURDOCK, G. P. (1959) *Africa: Its People and their Culture History.* New York.

17 LEWIS, I. M. (1961) *A Pastoral Democracy.* London.

LEWIS, I. M. (1969) 'Nationalism and particularism in Somalia'. In: P. H. Gulliver (ed.) *Tradition and Transition in East Africa: Studies of the Tribal Element in the Modern Era.* London.

18 HUNT, J. A. (1951) *A General Survey of Somaliland Protectorate.* Government Printer, Hargeisa.

19 LEWIS, I. M. (1960) 'The Somali conquest of the Horn of Africa'. *J. Afr. hist.,* **1,** 213–30.

20 MAFFI, M. (1960) *op. cit.*

21 CHOUMARA, R. (1960) *op. cit.*

22 See for example WOOD, A. M. (1969) 'The problems of communication for medical practice in East Africa'. *E. Afr. med. J.,* **46,** 536–40.

BURKITT, W. R. (1969) 'Rural mobile medicine in Kenya'. *Ibid.* 541–7.

COX, P. S. V. (1969) 'The value of mobile medicine'. *Ibid.,* 548–52.

23 LEWIS, I. M. (1961) *op. cit.*

24 BROWN, D. J. L. (1956) 'The Ethiopian–Somaliland frontier dispute'. *Intern. comp. Law Quart.,* **5,** 245–54, and (1961) 'Recent developments in the Ethiopian–Somaliland dispute'. *Ibid.,* **10,** 167–78.

DRYSDALE, J. G. (1964) *The Somali Dispute.* London.

HESS, R. (1966) *Italian Colonialism in Somalia.* Chicago.

LEWIS, I. M. (1965) *The Modern History of Somaliland.* London.

Somali Information Services (1962) *The Somali Peninsula.* Mogadishu.

25 BROWN, D. J. L. (1956 and 1961) *op. cit.*

26 MARRIAM, M. W. (1964) 'The background of the Ethio-Somalian dispute'. *J. mod. Afr. Stud.* **2,** 189–219.

27 CHAND, D. (1966) *op. cit.*

28 PROTHERO, R. M. (1965) *Migrants and malaria.* London.

29 For example, HUNTER, J. M. (1956) 'River blindness in Nangodi, Northern Ghana: a hypothesis of cyclical advance and retreat'. *Geogrl. Rev.,* **56,** 398–416.

9 Medical Geography and Health Planning in the United States: Prospects and Concepts

R. W. Armstrong

Traditionally, medical geography has concerned itself with problems of disease, problems which remain paramount in developing countries. Different problems are evident in the so-called developed nations. Here the major health concerns are those relating to the distribution and maintenance of health care services for a generally healthy population, to the care of increasing numbers and proportions of young and very old people, and to the protection and improvement of the health of the environment. During the 1960's in the United States notions of health planning received increasing attention. The endeavour offered new opportunity for medico-geographical contributions.

Health care in the United States currently faces a number of serious crises in cost, quality of care and equitable distribution of modes and standards of service to the population as a whole.[1] By all the usual statistical measures, health conditions compare unfavourably with Canada and many European countries.[2] Systematic planning to improve existing services and to implement new ones has been gradually gaining momentum under Federal government assistance. While current planning efforts remain uncoordinated and fragmentary, the general aim is to move from the present collection of independent, often unrelated health services and facilities toward what is envisaged as a *system* of health care.[3]

Health planning and health care in the United States has to be seen in its own context – a bewildering complexity of administrative structures, both public and private, amidst a great variety of health delivery systems. Medical care is delivered primarily through physicians in private practice, and by private hospitals with their own staffs. There is not the distinction between general practitioner and specialist as in Europe, and the majority of care is provided on a fee-for-service basis. Public health services, on the other hand, are provided by government agencies at federal, state, city and county levels.

All of these organisations have become involved in various ways with health planning. Some have been concerned with a specific problem, such as planning for hospital and medical facilities; some have a much broader focus as with community health on statewide, county or local levels. The chief stimulus for recent planning efforts has come through federal legislation, such as the Hill-Burton Act for hospital improvement and the Comprehensive Health Planning Act.[4]

COMPREHENSIVE HEALTH PLANNING

Comprehensive planning for the public's health is seen as a consortium of professionals, laymen and resources who must survey, analyse, plan, implement and evaluate their effort for their community within the boundaries of its geographical area. Federal funds are available for grants on submission of approved plans from states for statewide planning, and from health-related public and non-profit agencies within states for 'areawide' planning. The 'areas' are not specifically defined but in practice, for example, they have been a single county, a combination of two or more counties, or city wards. Planning councils are established at both the state and area-wide levels to initiate and implement planning and they must include representatives of government, health agencies and organisations, and a majority of representatives from the public.

Various interpretations have been given to the legislation which has started the processes of planning in many health agencies. In some cases it is seen as planning for comprehensive health services, that is, a streamlining and coordination of existing health facilities and services. In others a broader conceptual approach is followed where planning for the total health of man in his environment is envisaged. In these terms, comprehensive health planning should be prepared to improve health at any practicable point of intervention and not restrict itself to traditional spheres of action by health agencies.

This latter view is supported in the United States by a demonstrated need for new approaches to emerging health problems, such as those which involve human social behaviour and social organisation, and those involving the macro-environment, especially its pollution by the wastes of human activities. The concepts, techniques and organisation which have taken care of mosquito-ridden swamps and local sanitation problems, for example, are not so well adapted to tackling the social swamps of city ghetto and rural poverty or the politico-economic implications of country-wide environmental pollution control. Continuing health problems in communicable and non-communicable diseases, and in care of the aged, will remain a major part of health care, but future planning will emphasise, where possible, preventive rather than curative measures and seek more efficient modes of dispensing services.

CURRENT APPROACHES

Planning by health organisations in the United States refers to a variety of surveys, special studies and plans which may be short or long term. Few so far have achieved the sophistication of urban plans but useful techniques are being borrowed from this area.[5] Despite the attractions, in theory, of the systematic approaches in comprehensive health planning, it would appear that, in practice, individual innovative programmes have been more effective in making major changes. But the usefulness of alternative approaches notwithstanding, on-going health planning has become firmly established in the United States.

Health planning organisations generally seek four kinds of information about a community and its area. The first is geographical, although it is currently labelled *ecological*. This would include data on demography, housing, government, transportation, water supply and sanitation, and is usually rather superficial. The second is *prevalence* of all the known health problems. These data form the basis of establishing current patterns and future trends of disease, etc., and for deciding service needs and priorities. Thirdly, *utilisation* data are gathered to describe current use of existing health services and facilities. These would include statistics on hospital admissions, patient-origin and physician visits. Fourthly, the *resources* of the area are inventoried, mainly in terms of type of service, staff, organisation and cost.

THE NEGLECTED FACTOR OF LOCATION

A notable lack in most health planning is concern for locational factors, especially regarding resources. As Frieden and Peters have said, health plans are 'map shy'.[5] This is where medical geographers could make valuable contributions. The kind of geographical information readily available for local areas is usually inadequate for health planning purposes and has typically served only as padding for a general introduction to the planner's reports. There have been some attempts by community health service personnel to draw up survey schedules for gathering geographical data,[6] but they are useful mainly as factual checklists. There is a definite need for analytical interpretations of people and places which will help assess particular local health concerns and health services and which will aid in planning decisions best adapted to the unique conditions of each community and its habitat. In other words, there is a need for medical geographies.

Geographers have been making important contributions in one health-related field, namely studies on hospital utilisation, patient-origin, and location planning for new facilities.[7] Geographical survey methods, developed mainly in environmental health for appraisal of house quality and environmental hazards,[8] could be refined and extended to cover a wider range of health problems. Some health departments have begun to assess current problems, programme progress and future trends with the aid of on-going mapping surveys. These record week-by-week, for example, incidence of reported disease, sanitary complaints, accidents, child welfare visits and other statistics from all local agencies concerned with health and well-being. When coupled to a computer mapping system these programmes become effective surveillance tools.

Techniques and data currently used in these kinds of health surveys have limitations which require investigation. The deductive and descriptive techniques, such as mapping, are insufficient for most analyses of complex interrelations and must be supported in these instances by more inductive and analytical tools. Sampling schemes can be used to keep cost and time within reasonable limits. Many of the data now recorded are only symptomatic of the real problems; new forms of measurement more closely related to the causes of both old and new problems are needed.

SCALE CONSIDERATIONS

Geographers recognise and are well equipped to advise on questions of geographic or areal scale of data collection. Health planning requires working not only at the established administrative scales of nation, state, county and 'local', but also at a variety of other scales. Data for health in the United States have been collected and reported for decades at national and state scales, and to a lesser extent at the county level. Consequently, assessments of health problems on the basis of available data are most useful at national and state levels, only partially useful at county levels, and quite inadequate at local levels.

Public health and other agencies have also been structured for decades at national, state and county levels. These agencies naturally tend to view their health problems within the confines of their areas of jurisdiction. The state official sees state-wide problems and assesses priorities for planning accordingly; the county official works in similar fashion. Both are tied to their particular scale, and both are often unaware or unconcerned about neighbouring areas at the same or different scales. In addition, other agency personnel, and communities generally, have other divergent scales of operation.

Health problems assume different priorities at different scales, depending on the appropriateness of each level to handle them. At the local level, the problems which take first priority for handling are those which are local in scale and concern, such things as family health care, the specific care of infants and the aged, local social and welfare problems, water supply and sanitation. Problems on a larger scale, such as river pollution, or problems requiring special facilities and resources, such as mental illness, are better handled by state agencies, while effective air pollution control will necessitate national and perhaps global organisation. But it is the *priority* for dealing with a problem which varies at different levels of administration, not the *nature* of the problem. All problems in the health-environment system are interrelated and are not separate entities arranged in a convenient hierarchy to become the sole concern of one level of administration.

The geography of community health problems rarely conforms to the boundaries of administrative areas. Hence the basic value of the regional planning concept which recognises health problem areas and health service consumer areas. It also allows, where appropriate, for the coordination of existing health services to avoid unnecessary duplications and for the planning of new centralised facilities. This is revolutionary planning in the United States with its inflexible and separate administrative and legislative structures and where true regional planning on any major scale has yet to be embraced.

THE NEED FOR CONCEPTUAL FRAMEWORKS

In order that medical geography can contribute toward planning for improved public health, it must develop concepts within which relevant questions can be framed and examined. As Lukermann has so aptly stated: 'Geography is a catalogue of questions, and the questions – not the phenomena, not the facts, not the method – are geographic.'[9] Medical geography is concerned with questions about places, those areal complexes of man, culture and environment, in terms which are relevant to health. There are few conceptual approaches developed in the literature which seem applicable to the kinds of questions involving a community, its health and its habitat which are directly important for medical geographers working in health planning. One approach is to combine an ecological analysis to investigate the nature of health and environment interactions, and a chorological interpretation to provide the essential reference to place and areal associations.

LANDSCAPE EPIDEMIOLOGY

Pavlovsky and others in the Soviet Union,[10] and Audy in Malaya,[11] have used such an approach to the medical geography of certain zoonotic diseases, such as tularaemia, tick-borne encephalitides and scrub typhus. Pavlovsky's theory of 'landscape epidemiology' is based on a general ecological analysis of particular disease cycles examined in a number of sample field areas. The habitats and habits of the organisms involved – disease parasite, insect vector and animal hosts – are studied to identify parameters of the geographical environment that can be used as indicators (or 'ecological labels', to use Audy's term) of the relative potential of different areas to support the habitats of the disease cycles. A comparative geographical analysis is made using the indicators appropriate for a specific disease. These might be a number of particular measurements of terrain, vegetation, surface water, climate, animal populations, and human population and settlement.

ECOLOGICAL LABELS

Audy[11] suggests a somewhat different approach, namely, the identification of disease parasites as the 'ecological labels' of the assemblages of parasites and host animals which live in distinctive habitats. He recognises that different populations of the same host species have different disease patterns, and notes this as a reflection, in large part, of differences from place to place in the environment, including behavioural differences. These subtle geographical variations can be detected in the distribution of the 'ecological labels', thus pointing the way for a valid medico-geographical analysis.

These concepts were developed for communicable disease problems involving host-parasite relationships in essentially natural areas. Non-communicable disease problems of predominantly cultural environments require different approaches, but, with useful 'ecological labels' identified for chronic diseases, accidents and social health problems, new avenues of geographical analysis can be explored with a more likely prospect of leading toward explanatory models of the associations. Different approaches are also required when the focus of study is a community, as is usual in health planning, rather than the individual, or 'host', as in epidemiology.

A CONCEPTUAL MODEL

The conceptual approach outlined below was developed in order to carry out a medical geography of a small developed community with application to health planning. It adapts the general combination of ecological and geographical approaches used by Pavlovsky and Audy, together with ideas from Shimkin,[12] to the situation of a developed community. It proposes that the medical geography of communities and their habitats be approached through geographical concepts applied to the analysis and description of a simple health system comprising the community, its health concerns, and its health services.

GEOGRAPHICAL CONCEPTS

Geography is basically concerned with questions about place and the descriptions of places in some geographical order. Medical geography deals with medical phenomena in relation to place and seeks to identify the particular assemblages of that phenomena which distinguish one place from another and which are common to different places. It deals with assemblages, arrangements, and circulations of medical and related phenomena as systems of localisation. It recognises that the concept of place includes the historical development of all physical and cultural elements and their continuing emergence as new elements are added and the old are modified or disappear.[13] These fundamental geographical concepts must be included in research designs in medical geography.

A HEALTH SYSTEM MODEL

The general concept of health adopted here is that health is a state of adjustment between man and his environment at some point in time of the life sequence. The environment for man is both his external environment, or habitat, and his internal environment of body fluids. This ecological approach has been cogently presented by Sargent and Barr,[14] Dubos[15] and others. The immediate problem is to extract from the general concept a set of working propositions which can treat reality in manageable terms. For example, we may assume a health system that comprises: (1) the human community and its habitat, (2) the health concerns or problems which affect man's state of health, and (3) the modes of health care which the community has developed in response to health concerns. In one sense, the health concerns are the stimuli which initiate various kinds of response and as such they form a logical initial focus for study. They can be described and analysed in terms of their known aetiologies and ecologies, that is, the theories of process and interaction that would describe their role in causing human health problems. This first step then determines the directions of subsequent research on the community, its habitat, and its health services. Once the critical factors in the three components of the system have been isolated, attention can then be directed toward the workings of the system as a whole, its interactions and relative balances and imbalances. The system and its information flows can be represented thus:

HEALTH CONCERNS

The health concerns are, in the words of the poet Abraham Cowley, 'all the ills that men endure'. For the approach being presented, they are grouped into five broad categories of aetiology which may singly or in combination challenge states of health.

a. PHYSICAL (e.g. injury, surgery, malformations)
b. INFECTIOUS (e.g. viruses, bacteria)
c. CHEMICAL (e.g. nutritional, metabolic, tissue degenerative)
d. PSYCHOLOGICAL (e.g. mental illness)
e. SOCIAL-SOCIETAL (e.g. boredom, suicide, poverty)

Such a listing is subject to revision because knowledge is weak or lacking for many health concerns. The aetiology of cancer, for example, is unknown but the weight of current theory would classify this group of diseases as chemical. Nor are the categories mutually exclusive because most concerns combine elements of more than one. Any physical trauma, for example, always has its psychological impact. But the categories are useful because they encompass the entire range of known health concerns and conform to current theory and information on their origin, form and process.

THE COMMUNITY

In treating the community itself, the analysis follows lines which accommodate both current theories of disease aetiology and ecology and relevant published data. Five primary categories have proved to be useful:

a. HUMAN CHARACTERISTICS (e.g. age, race, work, diet, immunological status)
b. SOCIAL ORGANISATION (e.g. political, religious, social values, history)
c. HABITAT CHARACTERISTICS (e.g. location, climate, flora and fauna, housing, sanitation)
d. HABITAT ORGANISATION (e.g. land use, time-space values, location and design adaptations)
e. COMMUNITY CYBERNETICS (e.g. information flows, time-space relations, circulations, adaptive capacities).

These factors can be examined on the basis of the assemblage of health concerns known to affect the community under study. In practice, the aetiology and ecology of each concern serves as the justification for selecting and rejecting data on the community and for deciding the directions of valid analysis.

HEALTH SERVICES

Health services are treated under a three-fold categorisation:

a. MEDICAL CARE (e.g. preventive and clinical medicine, hospitals, insurance, infant care, elderly care, folk medicine)
b. ENVIRONMENTAL CONTROL (e.g. sanitation, housing, pollution control, safety)
c. BEHAVIOURAL INFLUENCE (e.g. health education, social work, psychiatric care)

The services are analysed in terms of the community they serve and the health concerns they deal with but not as an isolated component. They are an integral part of any particular health system and represent a response to health concerns and to health needs. Only when the services are examined in relation to the community and its health concerns can the adequacy of the responses be judged. Furthermore, the community, its concerns and its services form part of a web of larger communities, concerns and services which must also be taken into account. Many problems of the immediate community are shared by neighbouring communities. Many organised health services are provided by larger community administrations with regional rather than local purviews.

For any community under study, however, there will be a set of priorities of health concern that health services must handle. These priorities are the ones the community and health professionals recognise and it is upon them that the research of services can first focus with efficiency. For example, only a certain number of infectious diseases may be considered important, poor housing and high rents may underlie several health concerns, and solid waste disposal may be crude and dangerous. Subsequently, the analysis will assess, on the basis of current knowledge, the relative potential for specific health problems. The geography of the community and its habitat may indicate, for instance, most of the major conditions known to be necessary for an infectious disease or a social problem, one that is not yet evident because one or two key elements are missing. The nature of these missing elements may be identified and the likelihood of their appearance estimated.

RESEARCH DESIGN

The concepts just discussed can be incorporated into a research design with specific objectives. It should be emphasised that the health system concepts form the framework for posing geographical questions about location and relative location in terms appropriate to health planning. The health system concepts should not become the *tour de force* of the research design, but simply the vehicle. Emphases of the research analyses will vary according to objectives and to availability of data. For instance, more is usually known about the demographic characteristics and infectious disease problems of a community than about its cybernetics and psychological and social problems. Necessary sample surveys, case studies and other modes of collecting new data will thus be identified. Again, the chief thrust of the research may well be through, say, community behaviour. A thorough examination of the ways in which the community serves as actual and potential 'hosts' to health concerns may be more revealing than focusing initially on the concerns themselves.

Evaluation of the significance of health problems, community and habitat factors and health services in locational terms can be an important part of this kind of medical geography. Comparisons can be made on a distributional basis – problems are more serious in one place than in another, the community or its habitat varies from place to place, health services are available in some areas and not in others. Comparisons can be repeated for various parameters. We may judge health concerns on such bases as epidemiology, economic cost of handling the problem, the health services provisions necessary, and social values. Health problems take on different priorities according to the basis of their evaluation and such an assessment can form part of final interpretations.

CONCLUSION

Planning for improved health services in the United States is emerging from a piecemeal and confused initial period into what appears to be an established and more coordinated function. Medical geography adds strength to the locational aspects of all kinds of health planning, particularly in relation to location of resources, service areas and community assessment. The medical geography of a proposed planning area provides one sound basis for evaluating and establishing priorities. It gives a description of the assemblage of known health problems in an area, as related to the physical, biological and cultural spheres of the environment. This helps to identify those health and environmental factors which are area wide, those which are centred in sub-areas or segments of the community, and those which extend beyond the planning area. Study of the geographical associations of these factors thus reveals alternative points of intervention for handling known health problems and indicates where future problems might emerge. It also facilitiates demarcation of the areal extent of particular problems. In short, a medical geography provides one of the essentials needed for planning – a description of what places are like in terms useful to health planning.

REFERENCES

1 KNOWLES, J. H. (1970) 'U.S. health: do we face a catastrophe?'. *Look*, **34,** 74–8.
2 SHIMKIN, D. B. (1970) 'Man, ecology, and health'. *Archs envir. Hlth*, **20,** 111–27.
3 HANLON, J. J. (1969) 'An ecologic view of public health'. *Am. J. pub. Hlth*, **59,** 4–11.
4 United States Congress (1966) Public Law 89–749, November 3, 1966. Government Printing Office, Washington, D.C.
5 FRIEDEN, B. J. and PETERS, J. (1970) 'Urban planning and health services: opportunities for cooperation'. *J. Am. Inst. Planners*, **36,** 82–95.
6 Department of Community Medicine (1967) *Community Health Study Outline*. University of Kentucky College of Medicine, Lexington, Kentucky (mimeo).
7 MORRILL, R. L. and EARICKSON, R. J. (1968)'Hospital variation and patient travel distances'. *Inquiry*, **11,** 1–9.
8 American Public Health Association (1950) *An Appraisal Method for Measuring the Quality of Housing*. New York.
9 LUKERMAN, F. (1964) 'Geography as a formal intellectual discipline and the way in which it contributes to human knowledge'. *Can. Geogr.*, **8,** 167–72.
10 PAVLOVSKY, E. N. (1966) *Natural Nidality of Transmissible Diseases*. N. D. Levine (ed.). University of Illinois Press, Urbana, Illinois.
11 AUDY, J. R. (1958) Localization of disease with special reference to the zoonoses'. *Trans. R. Soc. trop. Med. Hyg.*, **52,** 308–28.
12 SHIMKIN, D. B. (1970) 'Space, congestion, crowding and isolation: preliminary analysis of problems and potential research'. Background document: *Man's Health and the Environment – Some Research Needs*. Report of the Task Force on Research Planning in Environmental Health Science. National Institute of Environmental Health Sciences, Washington, D.C., p. 242.
13 LUKERMAN, F. (1964) *op. cit.*
14 SARGENT, F. and BARR, D. M. (1965) 'Health and the fitness of the ecosystem'. In: *Environment and Man*. Traveller's Research Center, Hartford, pp. 28–46.
15 DUBOS, R. (1965) *Man Adapting* New Haven.

Part III
Spatial Definition

10 Atlases in Medical Geography 1950–70: A Review

A. T. A. Learmonth

INTRODUCTION

This review of atlases of disease aims to give some description but also some critical perspective in relation to the development of medical geography during the last twenty years or so. I have chosen to treat the topic under the headings of world atlases, atlases of developed countries and atlases from underdeveloped countries or regions, whether supra-national or intra-national.

WORLD ATLASES

The Welt Seuchen-Atlas is a monumental work by any standards,[1] and from time to time supplements, for example on cholera, are published by Jusatz and his colleagues from the Geomedical Research Unit of the Heidelberg Academy of Science.

There are 120 plates in the complete Atlas, many of which contain several maps. Thus Plate III/87 on plague pandemics of the twentieth century has a world map of the scale 1 : 45 million (about 720 miles to the inch), showing (1) by dated arrows the routes of dispersion of all plottable epidemic movements over the period, and (2) by shading areas with active endemic occurrence of plague to the present day, areas with rodent enzootic plague associated with the twentieth-century pandemic spread, and areas with permanent enzootic plague over the last few centuries. The reverse contains a world map, on which are appropriately placed monthly bar-graphs to bring out the differing patterns of seasonal incidence of plague in twenty-three countries; and there is also a bibliography of several hundred items. In Plate II/47, again, there are maps of the spread of individual epidemics across Europe 1899–1952, a set of graphs showing the numbers of plague cases by years for ten European countries, a map of the spread of plague by coastal shipping in the Eastern Mediterranean 1913–30, and again a large bibliography. Plate II/48 has a similar treatment of plague in Africa, 1899–1952, and Plate III/86 in the Americas 1899–1955.

Each such plate is accompanied by a substantial exposition by one or more of the contributors. These are analytical, at least at the level of subjective integration from very intensive map correlations and wide reading. Though the data are inherently quantitative, the analyses are seldom oriented towards modern statistical method.

As an example of the type of mapping and discussion I have chosen an extract from the large map of cholera in Asia in Plate III/81 (here reproduced as Fig. 10.1 in monochrome as against the original colour). Even

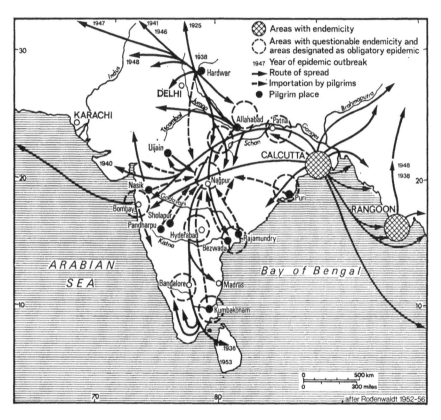

FIGURE 10.1 Part of Plate 111.8 (original multi coloured) from Rodenwalt, E. and Jusatz, H. J. (1952–61) *Welt Seuchen-Atlas*: Cholera in Asia 1931–55.

in translation this part of the text on the relations of rainfall and hydrography with cholera in Asia captures something of Rodenwaldt's style.

The multiplicity of ways in which precipitation or its absence can be of influence, shows how misleading a onesided evaluation of climatic factors can be. In their correlation with geological and hydrological conditions, with population density and patterns of habitation, they exert a contributing influence, which can be favourable or unfavourable.

L. Rogers has proved that the non-appearance of the east monsoon in south-east India has triggered severe epidemics in the area of the densely populated state of Madras (1877, 1892, 1900). Starvation as a consequence of the crop failure was joined with the diminution and pollution of the water sources.

In other regions, also in the endemic areas, such as Bengal and the river deltas of the east, the rain without doubt unleashes new epidemics, especially when the precipitation causes flooding and stagnation in these water courses. East of Allahabad, one approaches the centre of endemicity. Here the territory with frequent annual rains begins. In Bengal no time of the year is actually dry. And here dams and railway embankments increase the mass of stagnant water and consequent mortality from malaria, cholera and dysentery.

It is impressive to observe in North India the correlation between the amount of rainfall and the decline in epidemicity from east to west. Whereas in Bihar and parts of Uttar Pradesh the intensity of precipitation ranges from 1250 to 750 mm, at the altitude of Delhi it is only 650 mm. The figures are even more clear for the Punjab and all statistics show proof of the connection beween these figures and the mortality figures of epidemics.

Decisive, however, is not only the amount of precipitation, but what happens to the water. No doubt exists about the fact that stagnant water effects the promotion of cholera. It is different when large amounts of rain flow off quickly. On the basis of experiences in Java, Flu has emphasized that the onset of rainfall promotes the outbreak of cholera, whereas heavy continuation of rain brings such outbreaks to an end. Also, in India, torrential sweeping rains lessen the danger of cholera in the narrow and dirty streets of the villages.

In respect of the effect of precipitation, the geomorphological situation thus has a favourable or unfavourable influence, depending on whether the water is retained or runs off more or less quickly. Without a doubt water courses are pathways for the spread of cholera, both upstream and downstream.

Here is a definite correlation, which has not so far been the subject of an exact investigation, namely the correlation between the spread of cholera and the amount of water flow. In short, the reduction in the disease gradient of cholera stands in inverse correlation to the fall of gradient of the land and to the velocity of flow of the water. As already mentioned in respect to estuaries, rapid water movement is unfavourable to cholera.

Already the epidemiology of cholera in the great pandemics of the 19th century has shown that the disease avoided the upper reaches of the rivers, and all the more so as the streams approached the slopes of the mountains. This has always been the case in Germany on the northern slopes of the Alps, with their rapidly flowing streams, so also in Tyrol and in Switzerland, and, to cite an example much discussed in the literature, in the case of the alleged cholera immunity of Lyons, because of the location of this city on the swift flowing Rhone.

There is without a doubt a contrast between (a) the behaviour of the disease in areas of slowly flowing rivers with high organic material content, in areas of stagnant canals, lakes and ponds, and of deltas with water moving only slowly through ebb and flow and (b) its behaviour in areas of erosion, in which water flows rapidly with little organic material. . . .

It is almost self-evident, if the above-named correlation can be taken as valid, that a negative correlation between the spread of Cholera and altitude should also exist. This is supported by the fact that, as a rule, there is a lower density of population in the mountains. Even lower elevations, insofar as they are only thinly populated, can effect a hindrance to the influx of cholera. . . .

The incidence of cholera stands also in negative correlation to deserts and steppes. These also can be disease barriers. The Thar desert forms a protective wall in front of the southern part of Pakistan, seldom infringed. In North-West China, the farthest advances of cholera stop at the Gobi desert. But even deserts do not form insurmountable barriers when great masses of men are on the move in unusual ways (Mecca 1865), or when much travelled caravan routes serve as communication lines for the disease.

The *Welt Seuchen-Atlas*, then, is monumental, and inevitably had rather a limited circulation even among libraries. The text, descriptive and analytical at the level described, is very beautifully printed in the atlas in German and English, but of course retains the large format of the atlas itself (75 × 50 cm.).

May's *World Atlas of Diseases*, published with the *Geographical Review* 1950-5, is contrasted in several ways.[2] Overlapping maps between the two atlases are comparable in scholarship and cartography, but May's work is designedly lighter, wholly appropriate to its circulation with the *Review* to a very wide geographical public throughout the world and to nearly every library of repute. The *Review* carried short analytical articles, at a similar level of cartographic correlation and subjective integration to those of the *Welt Seuchen-Atlas*, but much of May's very substantial contribution to the literature of medical geography may be largely a consequence of the atlas project (see pp. 18–22).

May follows a flexible scheme of mapping and gives seventeen plates (95 × 63·5 cm.) of maps, mainly world maps. In some plates almost the whole sheet has been filled by a Breisemeister elliptical equal area map of the world, in others several world maps of this type but of smaller scale have been fitted in for comparison's sake or to show a range of different but cognate diseases; some use small inset world maps for comparison, and some have insets of crucial areas for the particular disease on a larger scale. Necessarily in a world atlas project, many of the diseases are shown by simple shadings or symbols indicating presence or absence of the disease, or sometimes its arthropod vector or vectors, or sometimes immunity in human populations as revealed by serological tests. Where the data permit, infection rates are shown, as in Plate 15 of yaws, bejel and pinta, in which the large map of the tropical lands includes material for simple cartographic correlation in mean annual isohyets. A monotone version of part of this map is included as Fig. 10.2. The second cholera plate has one map showing

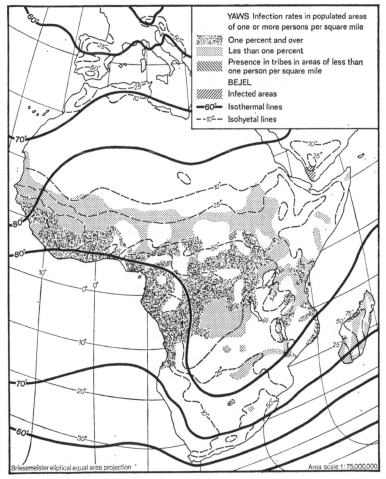

FIGURE 10.2 Part of Plate 15 from May, J. M. (1955) *The World Atlas of Diseases*: the world distribution of spirochetal diseases.

the spread of epidemic waves of cholera by coloured arrows, another with cross-classification of shadings for severity of infection rates and period of occurrence, a map of permanence of presence (endemicity) of cholera in southern and eastern Asia, small world maps of areas affected at important periods in the recent history of the disease, a map of Egypt showing by proportional symbols the number of cases in the 1947 epidemic, and one of India showing mortality in relation to humidity. Plate 14 of Leishmaniasis shows monthly incidence in relation to temperature and rainfall, the plate of poliomyelitis selects grades of infection at different periods. And so by these means the data and the problems are variously mapped.

DEVELOPED COUNTRIES

For the United Kingdom and the United States the field of disease-mapping using standardisation by age was opened up by Murray[3] (1962, 1967). Fig. 10.3 suggests the utility of deaths from all causes as socio-economic indicators. However for Britain his work was overlapped by the ambitious project for a *National Atlas of Disease Mortality*.[4] From this atlas I have selected as Fig. 10.4 the age-standardised map of cancer of the stomach in males for the period 1954–8. To quote Howe's commentary (in part):

> The outstanding feature of the maps of standardised mortality ratios for stomach cancer is the exceptional concentration of high ratios throughout the greater part of Wales. Ratios here are generally 10 to 60% above the national average. It should be appreciated, however, that vast tracts of mountain country in Wales are uninhabited, and elsewhere in areas above approximately 700 feet, the population is well scattered. In such areas mortality experiences are small and the calculated ratios are very often based on small numbers of deaths. Consequently, their reliability as a measure of underlying mortality may be affected. Even so, ratios are high in the densely populated industrial regions in the south and northeast of Wales. Stomach cancer accounted for 669 – or about 3·7% – of the 17,835 average annual male deaths in Wales.

Howe goes on to describe and comment upon each of five main areas in England with high standardised mortality ratios.

This may be compared with the map for the period 1959–63 (Fig. 10.5) and Howe's corresponding commentary. The reader will note the change of base map from a representation of the administrative units by area to symbols proportional to population placed approximately in the correct relative positions – a demographic rather than an areal base.

It is a pity that for the later period, or for both, both the areal and the demographic bases could not be used. Cartographic correlation with an-other map of a distribution pattern which is necessarily mapped by direct reduction of linear scale, for example, soils, becomes difficult. The benefits of areal-cum-statistical correlation using Chadwick's device,[5] of super-imposing a grid pattern on maps of distribution patterns for which the recording stations areas are *not* identical, are also lost. Correlation exercises were expressly excluded from Howe's project which was backed by a com-mittee of the Royal Geographical Society, including both geographers and

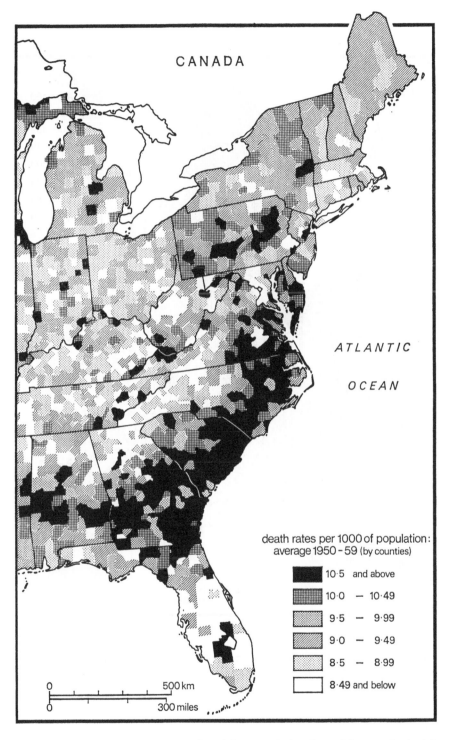

CANADA

ATLANTIC

OCEAN

death rates per 1000 of population:
average 1950 - 59 (by counties)

⬛	10·5 and above
▦	10·0 — 10·49
▩	9·5 — 9·99
▨	9·0 — 9·49
▫	8·5 — 8·99
☐	8·49 and below

0   500 km

0   300 miles

FIGURE 10.3 Part of Fig. 3 reproduced by permission from Murray, A. (1967)
Ann. Assoc. Am. Geogr., **57,** 307: Eastern United States: death rate variations from
all causes, average, 1950–9, age and sex adjusted.

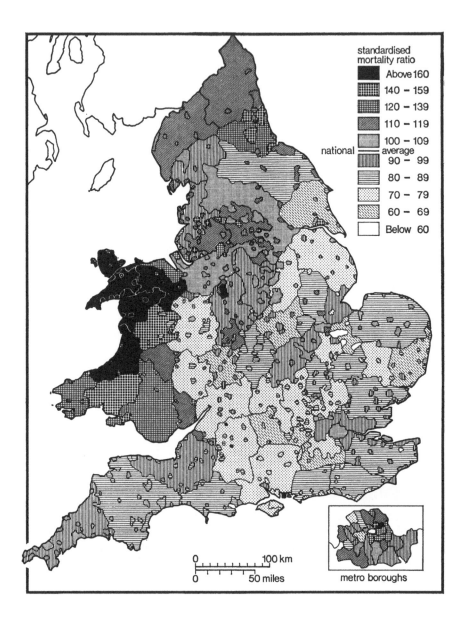

standardised
mortality ratio

Above 160
140 – 159
120 – 139
110 – 119
100 – 109
average
90 – 99
80 – 89
70 – 79
60 – 69
Below 60

national

0   100 km
0   50 miles

metro boroughs

FIGURE 10.4 From Howe, G. M. (1963, 1970) *National Atlas of Disease Mortality in the United Kingdom*, p. 35: cancer of the stomach (males) 1954–8.

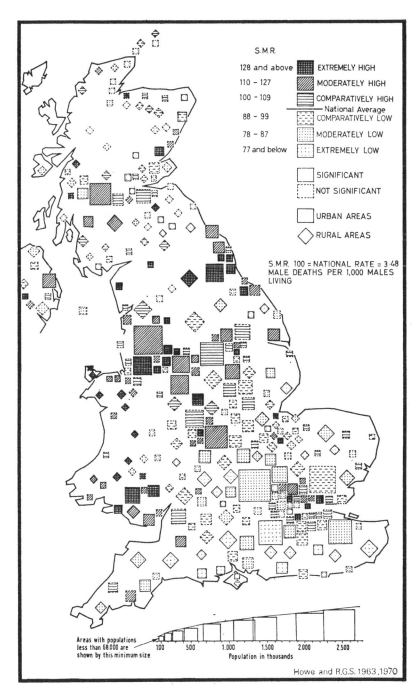

FIGURE 10.5 From Howe, G. M. (1963, 1970), p. 112: cancer of the stomach (males), 1959–63.

medical men. But Howe has since followed up certain problems in more detail, using computer graphics and more analytical techniques.[6] The later atlas uses dotted lines for areas *not* considered statistically significant at the 5 per cent level, as against continuous lines to delineate areas assessed as statistically significant (where a local standard mortality rate is more than twice the standard error from the national figure).

My two atlases (with collaborators)[7] of Australia's disease mortality do not attempt a similar assessment of statistical significance. On the other hand, the second atlas summarised in map form the areas of overlap between corresponding maps for the period 1959–63 and for 1964–8. These two atlases were compiled on a shoestring budget and rather quickly – the first to see if interest could be generated in medical geography in a vast country often said to have a monolithic culture and the second to compare periods round the 1961 and round the 1966 censuses. Indeed there may be a general case for quick and cheap atlas production, though no doubt by improved methods.

Multi-variate analysis ran beyond the time and man-power available but

> the study includes preliminary assays of cartographic correlations which seem suggestive, to see if more stringent correlation would be justified. None of the readily available distribution patterns which might be related to disease distribution, like rainfall, geological formations, industrial development, population density and the like have so far justified this form of analysis. . . .

> In interpreting choropleth maps of this type, it is not considered wise to attach significance to a single Statistical Division standing out by reason of high or low values, unless it is part of a gradient in the values. . . .

> In addition, in an attempt to show more general trends, it was decided to draw isopleth maps[8] to be used in conjunction with the choropleth maps. . . . These are valuable, we believe, in bringing out any spatial trends – hills and valleys as it were – from the data. In this respect they are easier to read than the choropleth maps. But it must be emphasised that, particularly where isopleths are based on large and sparsely populated Statistical Divisions, any particular isopleth represents simply a method of appreciating spatial trends based upon a certain cartographic convention.

This assumption of continuity of surface upsets some workers and even the editor of this book, well used to mapping conventions, has confessed himself exercised by the isomorts running across uninhabited desert. The method is illustrated (Fig. 10.6) in simply derived contours of mortality for cancer of the lung and bronchus (male) for Victoria for the period 1961–4; the approach has potential value by focusing attention separately on the generalised contours, just as in geomorphology from which the technique is borrowed, and then on areas of positive and of negative anomalies.

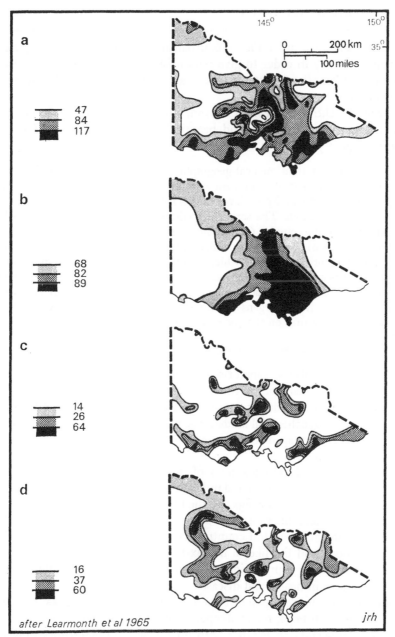

a

47
84
117

b

68
82
89

c

14
26
64

d

16
37
60

145° 150°

0   200 km 35°
0   100 miles

after Learmonth et al 1965

jrh

FIGURE 10.6 Generalised contours, cancer of the lung and bronchus (males), Victoria, Australia, 1961–4.
A. Standardised mortality ratios mapped as isopleths taking the geometrical centre of statistical divisions as 'spot-heights' and interpolating contours.
B. Generalised contours based on 'spot-heights' at the same points, each averaging the set of values falling within a circle of 10,000 sq. miles (26,000 km²) based upon each point; a map of regional trends.
C. Positive anomalies based on a comparison of maps A and B.
D. Negative anomalies based on a comparison of maps A and B.
(Source: A. T. A. Learmonth.)

In these atlases, and in Fig. 10.6, urban and rural data were not separately mapped, 'partly in order to give reasonable totals . . . though analysis of many problems is admittedly much stronger if they can be processed separately.'

From the U.S.S.R. the most relevant atlas is the medico-graphical section in the very comprehensive and beautifully produced *Atlas of Transbaykalia*.[10] Ignat'yev writes:

At the present time medical-geographic research in the Soviet pays much attention to the new development areas of Siberia and the Far East. Medical-geographic cartography is developing successfully on the basis of medical-geographic research. The series of medical-geographic maps that have been compiled for the complex *Geographical Atlas of Zabaikalia* are the first of their kind in the world. These maps reflect the medical-geographic character of the natural territorial complexes which are extremely complex in composition and possess a peculiar structure of territory, the Chita region and Buryat ASSR, which includes four physical-geographic regions. These maps are now widely used by local health organs for current and perspective planning of health protection measures on the 700,000 square kilometres of Zabaikalia.

This atlas comprises maps of:

(1) Curative places, sanatoria and rest homes, including information about the distribution of mineralised waters of various types.
(2) Biochemical endemics, goitre and renal calculus, related to several types of geochemical landscapes, mainly based on soil type, and with a separate map of the varying degree of risk of these endemics in the whole area.
(3) A large and complex map of medico-geographic regionalisation, including on the one hand indications of positive health in mountain climate, mineral springs and the like, and on the other hand liability to mountain illnesses (cold trauma, snow conjunctivitis and the like) and in the lower hill country tick-borne encephalitis, rickettsial infections and the like; a matrix-type key facilitates quick analytical or synthetic appreciation of the characteristics of the differing associations in health and disease.
(4) A companion map shows the distribution of vector ticks, classed in taiga, sub-taiga, forest steppe, steppe and tick-free areas.
(5) A large and detailed map with small insets of the territory shows the distribution of medical assistants, doctors, hospitals and hospital beds in relation to population density.
(6) Mammalian hosts of pathogens and carriers of disease, with main divisions into mammals of the steppe (voles, gopher-like suslik, marmots, pigmy shrew, hare, coypu etc.), and the modified groupings through forest steppe to mountain taiga and bare mountain

tops; limits are marked of important mammals – marmot, Brandt's vole, suslik, hedgehog and water vole, and cases of rabies from wild animals, and of tularaemia from coypu are plotted; (see Fig. 10.7), much simplified in order to produce a much smaller and monochrome map.

(7) Arthropod-borne infections with a valuable treatment of plague indicating the distribution and limits of marmot, or cases of human plague, and the varying regional prevalence of enzootic plague in rodents, and of outbreaks in the human population; associated zoonoses are also plotted – tick encephalitis, tick-borne rickettsial infections, tularaemia, rabies, tapeworms and the like.

It is difficult to think of the application of a similar amount of high quality of cartographic and printing skill to a regional atlas elsewhere, or of the allotment of such an important component to medico-geographical maps.

UNDERDEVELOPED COUNTRIES

Significantly perhaps, it is in this section that it is most difficult to distinguish between atlases proper and memoirs which use a whole series of maps of different diseases or other health indicators. Data are less accurate than those of developed countries and yet there is a lesser degree of generalisation than in many world maps, since it may be only one country that is being considered. So series of maps are used to give the most coherent picture possible of conditions of health and disease in the territory.

In South Asia there is more continuity than in many developed countries: insect-borne diseases enforced an ecological viewpoint. There was a pioneer piece of broad mapping and a valuable general discussion between the wars.[11] In a sense my own study of British India for the inter-war period was a complementary view by a geographer. Some workers interested in health in the tropics found a contribution in the use made of two-variable monochrome maps to represent together on one map both grades of average annual incidence and of variability in annual incidence; closeness of shading increased for higher incidence, direction of shading was altered according to three grades of variability – horizontal for low variability, oblique for medium and vertical for high variability.[12] The maps for nearly all the causes of death plottable from the natality and mortality data by administrative districts were complemented by some other maps, and a text using a good deal of the medical literature in an interpretative way. The aim was to attain a view of the regional burden of disease in relation to any other areal distribution patterns that seemed germane; to see regional health as a whole seemed an objective worth striving towards, at a

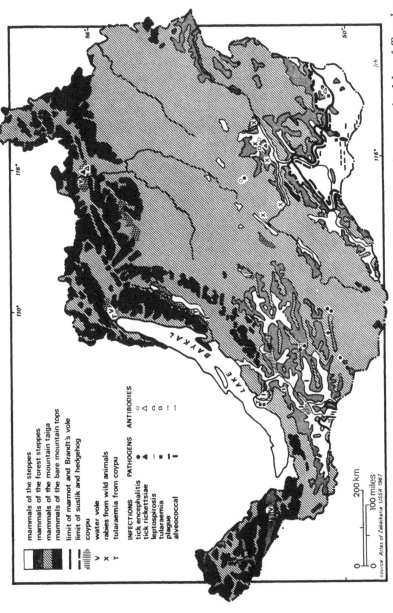

FIGURE 10.7 Distribution patterns selected for monochrome mapping from the colour plate, p. 116, *Atlas of Transbaykalya* (1967): mammals as hosts of pathogens and carriers of diseases.

The legend within the map reads:

mammals of the steppes
mammals of the forest steppes
mammals of the mountain taiga
mammals of the bare mountain tops
limit of marmot and Brandt's vole
limit of suslik and hedgehog

⩘ coypu
V water vole
X rabies from wild animals
T tularaemia from coypu

INFECTIONS   PATHOGENS   ANTIBODIES
tick encephalitis   ●   ○
tick rickettsiae   ◄   △
leptospirosis   ■   □
tularaemia   ■   □
plague   ▬   ▬
alveococcal   ▬   ▬

0   200 km
0   100 miles

source *Atlas of Zabaikalia USSR 1967*

jrh

time when the then recently independent countries were turning serious effort towards economic development and planning, national and regional. Some slightly more detailed mapping will be found in later work.[13]

Various parts of Africa, with serious problems of reliability and consistency of data, have nevertheless provoked the making of some very challenging atlases or atlas-based investigations.

The regional geo-medical memoirs on Libya and on Ethiopia discussed on pages 31–3 include what are really small general and health atlases. The general maps are included to give perspective, but some like 'Quality of Water' are specifically geo-medical. Cartography is of the high order we associate with the Heidelberg Academy of Sciences.[14] It is interesting too, that among textbooks in regional geography, one on Nigeria includes a modest but telling atlas, on rather a small scale, with a very useful discussion.[15]

Eastern and central Africa in the last few years have seen some remarkable developments in atlas presentation of admittedly imperfect data, showing both realism and ingenuity about the difficulties.

The nearest to a purely atlas project is the *Uganda Atlas of Disease Distribution*, though even here there are short but substantial analytical commentaries with each map.[16] This is a remarkable piece of collaboration between more than a score of medical men and a handful of statisticians and geographers. Selected diseases or topics covering groups of diseases (like 'other spirochaetal diseases') are treated in order according to the International Classification, 1965 Revision. The short commentaries are stimulating and authoritative and – unusually for an atlas – stand alone if even pioneer mapping a particular disease or disease group is impossible with the available data.

As an example of the cartography Fig. 10.8 concerns human trypanosomiasis or sleeping sickness.[17] The commentary (only part of which is quoted here) includes sections on the history, distribution, recent outbreaks, control and current research on the disease.

> The first recorded sleeping sickness epidemic in Uganda at the beginning of this century was a disaster for the people in the affected areas but resulted in a triumph for scientific medicine in the tropics. This was the discovery by Bruce, Castellani and Nabarro of the causal agent – a single-celled blood flagellate, the trypanosome – and its transmission by a species of tsetse fly which occurred in riverine and lacustrine habitats. This discovery suggested rational means of controlling and eradicating the disease and was of the greatest practical importance for people in all areas of Africa where the disease occurred or can occur. . . .
>
> *Distribution*
> The occurrence and course of sleeping sickness in Uganda from 1900 to 1960 is illustrated in [Fig. 10.8]. These outbreaks have been adequately described

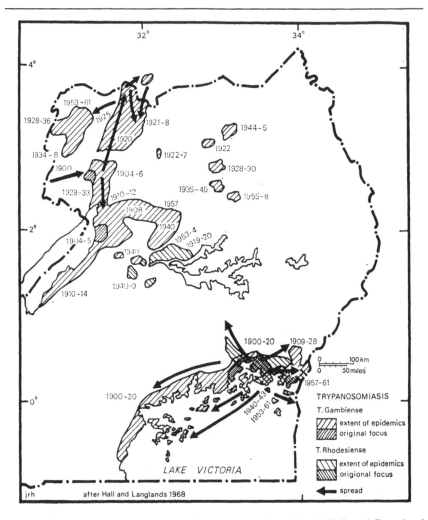

FIGURE 10.8 From Map 1, p. 62, Onyango, R. J. (1968) *in* Hall and Langland (eds.) *Uganda Atlas of Disease Distribution*: sleeping sickness (human trypansomiasis) Uganda 1900–60.

in a series of papers by Morris (1959, 1960, 1962), respectively relating to Lango, western Uganda (Lake Edward and George), Bunyoro and the Albert Nile; by McKichen (1944) on the first outbreak of *T. rhodesiense*, sleeping sickness, and by Robertson and Baker (1958) and Robertson (1963) both relating to epidemiological features of the disease along the North-eastern shore of Lake Victoria. Other details of the incidence of *T. gambiense* and *T. rhodesiense* infections, and of the foci in Uganda are given in the Annual Reports of the Tsetse Control Division of the Department of Veterinary Services and Animal Industry of Uganda.

Recent Outbreaks

Since 1960 the two main foci of sleeping sickness have been in south-eastern Uganda (Busoga and Bukedi Districts) and in West Nile District. The second map (not illustrated) shows the distribution and annual incidence of cases confirmed by blood slide examination and notified to the Ministry of Health.

Langlands contributes an interesting concluding essay on 'The geographic basis to the pattern of diseases in Uganda'.[18] He reviews latitude, open water, swamps, temperature, rainfall (mapping the average number of dry months), vegetation, and in human geography particularly, migration and immigrant groups, urbanisation, food crops, economics and population density.

In central Africa our editor's doctoral thesis was based on an atlas (see pp. 199-210 in Chapter 15), but reaching much further into analytical work on the frontier of the discipline.[19] His own summary is:

A method of surveying spatial patterns of morbidity suitable for use in underdeveloped countries is described and applied to a large area of central Africa. In conjunction with a simple computer graphical programme, the method is used to compile 55 maps of individual diseases and 19 maps of environmental or hospital factors which might influence disease. New associations which emerge from these geographical comparisons are offered as tentative hypotheses for testing in other disciplines.

Thus he wrote of relationships between cancer of the oesophagus and locally varying preparations of alcoholic spirits – a topic which he has developed since (see pp. 247-57). This section included assessments of evidence by a botanist and a chemist; the author is careful not to claim more than a geographer's competence, but the geographer's contribution is clear. All positive or negative relationships are assessed in relation to the indications they give for further work. Thus a spatial relationship between diabetes mellitus and cassava as a staple food (rather than maize) might be worth exploring,[20] but cannabis addiction is less likely to justify further investigation as a means of preventing cancer of the thyroid glands. A longer discussion[21] suggests some reappraisal of the geographical evidence for Burkitt's tumour to contrast the original east African data with tumour experience in central Africa. McGlashan confirms an inverse association with altitude, but points out that tumour incidence tends not to vary significantly with mean temperature in the coldest month. Thus the atlas work is a basis for analysis, in part to produce new hypotheses worthy of further investigation, in part to urge reassessment of existing hypotheses.

Burkitt has since collaborated with a geographer in a further study of the geography of cancers in East Africa;[22] the memoir contains a very substantial section assessing the need for study of cancers in the area chosen (Kenya, Tanzania and Uganda) and the methods of collecting data, necessarily relatively inaccurate, and of off-setting the inaccuracy as far as possible, and yet it is not too fanciful to treat it as an atlas-based project. A very useful introductory section contains maps of hospitals, of relief, of settlement potential (combining mapping of population densities and of rainfall probabilities, generalised from classic sources on the area). Then cancer of the oesophagus for men is mapped both by hospital and by tribe, by proportional symbols, followed by maps by hospitals for cancer of the penis, Kaposi's sarcoma, cancer of the liver, cancer of the stomach, scar epithelioma (on the site of old tropical ulcers of the skin), and (for both sexes together) for Burkitt's lymphoma. Tentative conclusions invite further investigations rather than solving a problem, but specific pointers are offered. Thus to follow up Burkitt's tumour:

> Much has been written about the distribution of this tumour. The geographical pattern was first studied in East Africa and it was the geographical variations exhibited by this tumour that initiated the study of other tumours by the methods described above. The information depicted indicates the high incidence in the warm and relatively moist regions round Uganda and in the coastal regions of Kenya and Tanzania. It also illustrates the rarity of the tumour in the highlands of south and west Uganda, central Kenya and northern Tanzania. The rarity in central Tanzania is believed to be related to the long dry season experienced annually there.

> It is now believed that the climatic dependence demonstrated for this tumour indicates not, as previously postulated, a relationship to some insect vectored virus, but a relationship to hyper- or holo-endemic malaria, a relationship that has now been demonstrated geographically on a world wide basis (Burkitt 1969) and in detail in Uganda (Kafuko et al. 1969).*

> In Uganda, the haemoglobin electrophoretic pattern of patients with Burkitt's lymphoma was compared with that of non-related controls of the same age and sex from the same village to see whether there is a higher proportion with sickle cell anaemia among the controls (Pike et al. 1970).† Initial results suggest that children with AA genotype appear to have approximately twice the chance of developing the tumour as those with AS genotype. In view of the established negative relationship between severe malaria and the sickling trait, a similar relationship with Burkitt's lymphoma would lend support to the geographical hypothesis of an association between Burkitt's lymphoma and malaria.

> One of the most interesting features of the distribution of Burkitt's lymphoma, in view of the possible involvement of an infectious agent, is the observation in the West Nile district of Uganda of a geographical drift of cases with time

* Burkitt, D. P. (1969) *J. nat. Cancer Inst.*, **42**, 19. Kafuko, G. W., Bainganga, M. and Tibemanya, J. (1969) *E. Afr. med. J.*, **46**, 1.

† Pike, M. C., Morrow, R. H., Kisuule, A. and Mafigiri, J. (1970) *Br. J. prev. soc. Med.*, **24**, 39.

(Pike *et al.* 1967, Williams *et al.* 1969).* When the cases occurring in six years were mapped so that the symbol for each patient indicated both the place of residence and the date of onset, it was clear that year by year the area of maximum occurrence had shifted across the district, with a return in the final year to the initial centre of concentration. The distance and time between patients is generally so large as to preclude any possibility of case to case transmission but suggests an infection to which all are exposed but which leads to cancer in only a small minority of cases.

CONCLUSION

This survey of atlases in medical geography must inevitably be incomplete because of barriers of language and accessibility. A more complete survey might well start from the fine bibliography from the Leipsig Deutschen Instituts für Länderkunde.[23] Even so one may discern some broad patterns of development from this paper. The world atlases described have each played their own unique part in the development of post-war medical geography. There may well be a place for renewed effort in this sphere, perhaps in a relatively cheap and compact form which is able to be revised frequently. This might be a major service in relation to the problems of exotic disease mentioned on p. 30 of this book. The atlases from developed countries are clearly full of promise, especially if they become available in time series coinciding with censuses, and are complemented by the deeper analytical studies in depth, of which some forerunners are already in existence. Rather against expectations, the underdeveloped countries, especially in Africa in recent years, have provided studies which show how atlas-based studies can be analytical and hypothesis-finding, without in any way outrunning the content proper to an atlas or the competence of an atlas-maker or atlas-making team.

REFERENCES

1 RODENWALDT, E. and JUSATZ, H. J. (1952-61) *World Atlas of Epidemic Diseases*. Hamburg.
2 MAY, J. M. *World Atlas of Diseases*. American Geographical Society, New York. Also published with short descriptive and analytical texts in the *Geographical Review* (1950) **40** to (1955) **45**.
3 MURRAY, M. (1962) *Ann. Assoc. Am. Geogr.*, **52**, 130.
   MURRAY, M. (1967) *Ann. Assoc. Am. Geogr.*, **57**, 301.
4 HOWE, G. M. (1963, 1970a) *National Atlas of Disease Mortality in the United Kingdom*. London.
5 CHADWICK, J. G. (1961) *Geography*, **46**, 25.

* Pike, M. C., Williams, E H. and Wright, B. (1967) *Br. med. J.*, **2**, 395. Williams, E. H., Spit, P. and Pike, M. C. (1969) *Br. J. Cancer*, **23**, 235.

6 HOWE, G. M. (1969) *Nature*, **223**, 891.
  HOWE, G. M. (1970b) *Spectrum*, **71**, 5.
  HOWE, G. M. (1970c) *R. Soc. Hlth J.*, **90**, 16.
  REID, JEAN (1970) *Times Educ. Suppl.*, 1 May.

7 LEARMONTH, A. T. A. and NICHOLS, CHRISTINE, G. (1965) Maps of Some Standardised Mortality Ratios for Australia 1959–63, and LEARMONTH, A. T. A. and GRAU, R. (1969) Maps of Some Standardised Mortality Ratios for Australia 1964–8. Occasional Papers Nos. 3 and 8 Department of Geography, Australian National University School of General Studies, Canberra.

8 MACKAY, J. R. (1951) *Economic Geography*, **27**, 1.

9 WELLS, R. and KUPKEE, L. (1966) *The Safety Zones: Differential Mortality Rates in Australia 1961–3*. Commonwealth Department of Health, Canberra.

10 *Geographical Atlas of Transbaykalia* (1967) Academy of Sciences, Siberian Division, Moscow/Irkutsk.
  IGNATYEV, YE. I. (1968) 'Medical geography in practice'. In *Melanges de Geographie physique, humaine, economique, appliquee*, offerts à M. Omer Tulippe. Gembloux.

11 MEGAW, J. W. D. (1927) *Indian med. Gaz.*, **42**, 299.

12 LEARMONTH, A. T. A. (1954) *Liverpool Ann. trop. Med. Parasit.*, **48**, 354.
  LEARMONTH, A. T. A. (1958) *Indian geogr. J.*, **33**, 1.
  LEARMONTH, A. T. A. and PAL, M. N. (1959) *Erdkunde*, **13**, 145.

13 LEARMONTH, A. T. A. (1961) *Geogrl. J.*, **127**, 10.

14 KANTER, H. (1967) *Libya* and FISCHER, L. (1968) *Afghanistan*, Springer-Verlag Geomedical Monograph Series Nos. 1 and 2: ed. H. J. Jusatz. Berlin–Heidelberg–New York.

15 BUCHANAN, K. M. and PUGH, J. C. (1955) *Land and People in Nigeria: the Human Geography of Nigeria and its Environmental Background*. London.

16 HALL, S. A. and LANGLANDS, B. W. (eds.) (1968) *Uganda Atlas of Disease Distribution*. Department of Preventive Medicine and Department of Geography (Occasional Paper No. 12), Makerere University College, Kampala.

17 ONYANGO, R. J. (1968) 'African human trypanosomiasis', p. 61, Hall and Langlands, (1968) *op. cit.*

18 LANGLANDS, B. W. (1968) 'The geographic base to the pattern of disease in Uganda', p. 176, Hall and Langlands (1968) *op. cit.*

19 MCGLASHAN, N. D. (1968) *The Distribution of Certain Diseases in Central Africa: an approach to medical geography in an underdeveloped country*. University of London Ph.D. Thesis.

20 DAVIDSON, J. C., MCGLASHAN, N. D., NIGHTINGALE, EVELYN and UPADHYAY, J. (1969) *Medical Proceedings Johannesburg*, **15**, 426.

21 MCGLASHAN, N. D. (1969) *Internat. J. Cancer*, **4**, 113.

22 COOK, PAULA and BURKITT, D. (1970) *An Epidemiological Study of Seven Malignant Tumours in East Africa*. Medical Research Council, London.

23 VON HELMUT ARNHOLD (1968) *Medizinische Geographie – eine Auswahlbibliographie*. Sonderdruck aus Wissenschaftliche Veroffentlichungen des Deutschen Instituts für Länderkunde, Neue Folge 25/26.

11 Blindness in Luapula Province, Zambia

N. D. McGlashan

PHYSICAL ENVIRONMENT

The Luapula Province of Zambia (formerly Northern Rhodesia) lies about
10° south of the Equator and midway east to west in the heart of Africa
(see Fig. 11.1). The province includes in its eastern sections a part of the
high plateau of north-eastern Zambia and, to the west, all that part of the
Luapula River lowlands on the east bank and to the east and south-east
of Lake Mweru. The river and the middle of the lake form the boundary
with the Congo Republic (Kinshasa). This plateau or valley dichotomy is a
dominant feature of the human geography as well as the geomorphology
of the Province.

The Miocene drainage of the plateau, now at 4,400 ft., was towards the
south-south-west.[1] Upon this surface the proto-Mweru and proto-
Bangweulu lakes developed either by erosion or by crustal warping until,
in late tertiary times, two great drainage captures occurred. The Luvua
tributary of the Congo cut back into Lake Mweru[2] and the Luapula cut
back to capture the Chambeshi–Bangweulu outflows. Rejuvenation nicks
in the Luapula course show waterfalls above Lake Mweru and below Lake
Bangweulu.

On the plateau, stream courses are sluggishly mature and often actively
aggrading their beds, particularly in the Bangweulu Swamps area.[3] The
rivers of the plateau edge show attributes of youth falling steeply 1,200 ft.
or so to the Luapula valley, where the main river meanders northwards in
broad curves across a wide and densely populated alluvial plain.

The peoples of the Luapula seem all to have come in at least four invasion
waves of broad ethnic similarity from a heartland empire referred to as
'Kola' situated to the west, possibly near the present Angola–Congo–
Zambia frontiers.[4] From 1800 to 1900 the slave-raiding activities of
Swahili, Arab and Yeke weakened the tribes and paved the way for the
coming of the Pax Britannica (Fig. 11.1).

That peace has been responsible for a startling population growth –
estimated at fourfold – between 1900 and 1960.[5] This has made the Luapula one of the most densely populated parts of rural Zambia today.[6]

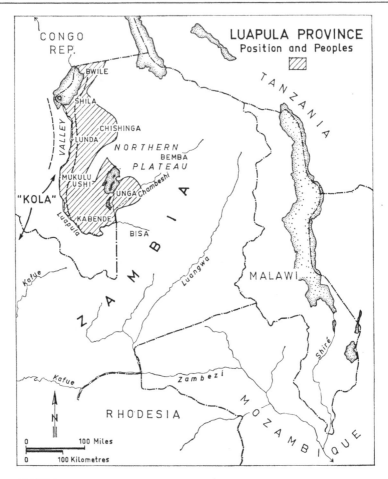

FIGURE 11.1

This increase of population has thrown great stress upon the natural resources of the province. The plateau soils are low in inherent fertility, being sandy-textured, strongly leached and underlain by laterite. In the valleys of major rivers an increased humic content renders soils marginally more fertile.[7]

The most widespread vegetal type is light woodland with intervening 'dambo' areas of open coarse grassland. Major game animals used to abound at the turn of the century but now small buck and wild pig are all that remain. Tse-tse fly infestation is still a deterrent to domestic stock in most of the province, and indeed supply of an adequate diet is a widespread problem.

For this reason fishing – the province's greatest industry – is doubly important. The sale of fish, almost entirely to the Copperbelt by road, is the greatest earner of cash and also fish eaten locally is the only substantial source of protein in an otherwise largely carbohydrate diet. Cassava (manioc) is everywhere the dietary staple.

HISTORY OF BLINDNESS

Before 1900 records of literate visitors are few. The earliest Europeans to reach the river were Dr. Lacerda and his party who visited Chief Kazembe of the east Lunda in 1798.[8] A century later Crawford travelled in the area in 1893, 1897 and 1911.[9] Neither of these record any mention of blindness. Lammond, speaking from personal knowledge as a resident from 1902 onwards, has stated that blindness was unusual up to 1914. Indeed early administrative records do not start to mention blindness until 1934, even though measles outbreaks started to occur with frequency from 1921.[10] A sharp rise in blindness seems rapidly to have occurred for, by 1935 and 1936, the need was so great that Blind Schools were set up by the Christian Mission in Many Lands both at Luela (211, see Fig. 11.2) and at Johnston Falls (127). The war, 1939–45, caused an acute shortage of both administrative and medical personnel and no further action is recorded, until, in 1946, a special investigation was recommended by a (British) Colonial Office Blind Welfare delegation visiting the province.

During the next decade alarm increased and rates of blindness of 25 to 30 per 1,000 persons were estimated, with particular emphasis on high incidence among children. A practical step in 1956, with results subsequently confirmed, was an entomological search which found no simulium flies, which are the vectors of river blindness. This disease is blessedly absent from the Luapula Province.*

In the same year the ophthalmologist, C. Malcolm Phillips, made a detailed field tour concentrating attention on Chief Mununga's area (103 and 104 on Figs. 11.2 and 11.3). His report[11] quotes an overall rate of 23·7 for all ages, which bore out the earlier estimated figures. His other main statistical findings were that, in 80 per cent of cases seen, measles was associated with the onset; that over 80 per cent of the blind went blind before the age of ten years and that 'it can be definitely stated that no signs of malnutrition were seen'. Phillips also concluded that much unnecessary damage to eyes was being caused by customary African remedies, and he was able to quote many instances of this.

BLINDNESS DATA

Phillips' report, together with offers of assistance and continuing pressure from the Royal Commonwealth Society for the Blind (R.C.S.B.), led to the

* See also Chapter 19, p. 261.

passing of the Northern Rhodesia Blind Persons' Ordinance, Cap. 207 of 1961. By its provisions, Blind Boards in the provinces were given decentralised authority to register every blind person in the territory. The provincial registers included particulars of the patient's tribe, home, age, sex, cause and date of onset of blindness and details of capacity to profit from training facilities. A practical field definition of blindness was to be used. 'A person who by reason of impaired sight is unable accurately to count separated fingers held at a distance of six feet from his face while both eyes are open shall be considered blind.'[12] For the Luapula Province only, the R.C.S.B. made available a Mobile Eye Clinic for 1962–4 which was to carry out educational and remedial work as well as actively register blind persons. The clinic was based at Nchelenge (105) near the worst affected area and was in the permanent charge of an experienced male nursing officer,

TABLE 11.1

Enumeration Area	Area sq. ml.	Population	Male Blindness	Female Blindness	Gross Blindness Rate Blind/1000
101 Puta North	94	3,537	11	9	5·65
102 Puta South	484	6,009	15	19	5·66
103 Mununga North	215	11,110	49	33	7·38
104 Mununga South	413	8,633	8	11	2·20
105 Kambwali North	156	9,311	22	29	5·48
106 Kambwali South	108	4,866	11	13	4·93
107 Kanyembo North	193	6,319	16	10	4·11
108 Kanyembo South	95	5,553	13	8	3·78
109 Kawambwa	53	6,190	4	8	1·94
110 Munkanta	1,060	8,476	7	6	1·53
111 Mushota North	245	2,411	3	3	2·49
112 Mushota South	376	4,583	4	5	1·90
113 Chama North	856	8,327	16	13	3·48
114 Chama South	535	1,976	1	–	0·51
115 Kazembe North	67	15,025	18	16	2·26
116 Kazembe Central	157	1,223	7	4	8·99
117 Chisenga Island	30	2,161	1	1	0·93
118 Kazembe South	190	7,215	13	4	2·36
119 Mwenda West	384	1,131	–	1	0·88
120 Mwenda East	918	5,362	3	4	1·31
121 Mutipula	550	5,883	4	7	1·87
122 Lukwesa North	130	7,676	3	2	0·65
123 Lukwesa South	76	7,061	8	5	1·84
124 Lubunda North	121	6,324	8	8	2·53
125 Lubunda South	115	4,785	5	1	1·25
126 Kashiba North	74	4,000	15	7	5·50
127 Kashiba South	109	4,358	7	4	2·52
128 Mulunda North	79	5,285	4	5	1·70
129 Mulunda South	118	2,699	2	3	1·85
130 Katuta Kampemba	375	4,297	7	1	1·86
Game and Waste	920				
Kawambwa District	9,296	171,786	285	240	3·06

TABLE 11.1 (contd.)

Enumeration Area	Area sq. ml.	Population	Male Blindness	Female Blindness	Gross Blindness Rate Blind/1000
201 Ft. Rosebery	Urban	4,927	–	–	–
202 Chimese East	178	4,191	4	1	1·19
203 Chimese West	200	7,841	12	3	1·91
204 Matanda North	509	3,573	1	1	0·56
205 Matanda South	693	5,134	–	–	–
206 Kalasa Lukangaba	411	8,538	5	4	1·05
207 Kalaba	524	8,631	7	2	1·04
208 Chisunka	308	5,960	1	3	0·67
209 Mabumba	533	9,123	9	10	2·08
210 Mibenge	216	7,277	3	4	0·96
211 Kasoma Lwela	520	6,137	1	1	0·32
212 Milambo	1,683	9,721	5	7	1·23
213 Sokontwe	459	5,106	1	–	0·20
Ft. Rosebery District	6,234	86,158	49	36	0·99
301 Mwansakombe West	84	5,259	3	5	1·52
302 Mwansakombe East	138	7,140	4	–	0·56
303 Mwewa North	158	3,941	5	1	1·52
304 Mwewa East	164	9,106	10	5	1·65
305 Mwewa West	107	1,022	1	–	0·98
306 Chitembo	71	9,794	3	8	1.12
307 Chishi Island	23	5,142	1	–	0·21
308 Mbabala Island	44	3,431	–	–	–
309 Kalimankonde	760	4,020	–	–	–
310 Kasoma	160	5,152	–	–	–
311 Nsamba North	117	3,118	–	1	0·32
312 Nsamba South	133	2,684	1	–	0·38
313 Bwalya Mponda North	165	3,274	–	–	–
314 Bwalya Mponda South	266	1,812	–	–	–
315 Samfya Town	Urban	4,145	3	–	0·72
316 Kasoma Bangweulu W.	103	2,964	5	5	3·37
317 Kasoma Bangweulu E.	215	6,002	3	3	1·00
318 Kalasa Mukoso North	235	5,070	1	–	0·20
319 Mulakwa	75	10,595	4	3	0·65
320 Kalasa Mukoso South	970	7,274	1	–	0·14
Samfya District	3,988	100,945	45	31	0·75
LUAPULA PROVINCE	19,518	358,889	379	307	1·91

P. C. G. Adams. During two dry seasons he was visited and accompanied by ophthalmologists from St. Thomas' Hospital, London: B. Cobb[13] in 1962 and P. Awdry[14] in 1963.

Whilst the Luapula register of blind persons provided the best available numerical information by the time of its completion in October 1964, certain limitations should be noted.

Three groups of blind persons *may* not have been completely registered. Firstly some blind persons may never have been located, especially perhaps, in the very difficult terrain of the Bangweulu swamps (309–14). As registration proceeded, the cost of recording each new case increased with the difficulty of finding new blind persons and this search eventually reached an uneconomic level. At that time probably under 7 per cent[15] of blind persons remained undiscovered and these would be likely to be atypically old and probably unable to benefit from Blind Board training facilities.

Secondly some 107 persons, including only two children, refused to give particulars for registration. The reason would have been that rural people may see no personal advantage in answering intrusive personal questions.

Thirdly some blind persons from the Luapula living elsewhere, perhaps as beggars in the towns for instance, may have given other home addresses and have been registered in other provinces. Indeed the Luapula register itself included seventy-eight patients known to be from the neighbouring Northern Province. These have been excluded from this geographical survey.

Lastly it was theoretically possible for a blind person to be removed from the register in case either of death or of recovery of sight, possibly after a successful cataract operation. In no case, up to October 1964, had that procedure been followed, although two successful cataract operations were known and surely some blind persons must have died in a period of two and a half years.

An inaccuracy of another kind stems from retrospective medical diagnosis of cause of blindness when the actual blinding lesion may be many years old. Often non-medical circumstantial evidence may be all that is available. Age of onset too is only approximately recorded in most cases.

Whilst registration was proceeding, the first full census of Northern Rhodesia was taken in May and June 1963. Sex was recorded and three divisions only of age: born before 1918, born before 1942 and born after 1942.

Table 11.1 brings together the final Blind Register figures and the census information by census enumeration areas. These were usually a clearly defined half of a Chief's area (e.g. 101 and 102) and had an *average* population in this province of about 6,000. Since the purpose was to define variations of geographical incidence, rates were calculated for each enumeration area and quoted per thousand of population. The usual convention of quoting rates per hundred thousand was not followed because it seemed to imply a measure of spatial uniformity. In fact the study tended to emphasise spatial variation at a local level. The differences of rates between the three districts of the province are themselves a clear example of this feature (see Table 11.1 district sub-totals).

In addition to Census and Blindness information the third source necessary was base maps. The province is fully covered at a scale of 1:250,000 on a transverse Mercator projection. On these maps, villages rarely appear and are never named so that the exercise of mapping the blind (Fig. 11.2) required first of all a full map of all villages in the Province. Fortunately most of this enormous labour had already been carried out by Kay[16] in 1961 and 1962 working from administrative tour reports. In 1963 it was merely necessary to re-locate any villages which had moved or changed their Headman's name. Additional confusion arose by certain names being common both within and between Chief's areas.

BLINDNESS MAPPING

The dot map of the distribution of all blind persons (Fig. 11.2) emphasises the overall preponderance of the blind in the low-lying valleys close to the Luapula river and its extension along Lake Mweru's shores. In fact, the line of villages along the valley road and its deviations back from the river to avoid swampy places can actually be followed in the blindness pattern. This is merely a reflection of the settlement pattern (e.g. on the border of areas 108–15).

Negatively the map shows the whole area of the swamps (309–14) having only two blind persons. This may well indicate under-reporting, especially when it is contrasted with the apparently comparable area on the western shore of Lake Bangweulu, where the predominantly fishing populations of Chiefs Mwewa and Chitembo (304 and 306) record quite high numbers of blind persons.

Turning to the plateau areas of Kawambwa (109–14 and 119–21) and Fort Rosebery (201–13), the scatter of blind persons seems light with the exception of a bad area near Chief Chama's court (113). Each dot, however, is an afflicted human being and the details of many of the causes[17] show this to be avoidable blindness and suffering that will be borne for a whole lifetime.

Blindness rates per thousand of population are shown in Fig. 11.3, where the class intervals were selected by scatter diagram to emphasise discontinuities in the distribution. The pattern shows a high incidence of blindness, both relatively and absolutely, in the remote areas of the north. Some of these areas (101–3, 111 and 113) are extremely difficult to reach by road, especially during the rainy months from mid-October to April.

The line of riverine enumeration areas from Mulunda (129) northwards to Mununga's (103) shows, with minor variations, mainly high rates of blindness such as first drew attention to the local problem.

In the two southern districts, on the other hand, inaccessible or remote areas do not show high blindness rates. Matanda South (205) is very difficult to reach by road but reports no blind persons. The swamps too (309–14), which are only accessible by water, record only two blind persons amongst 20,000 inhabitants.

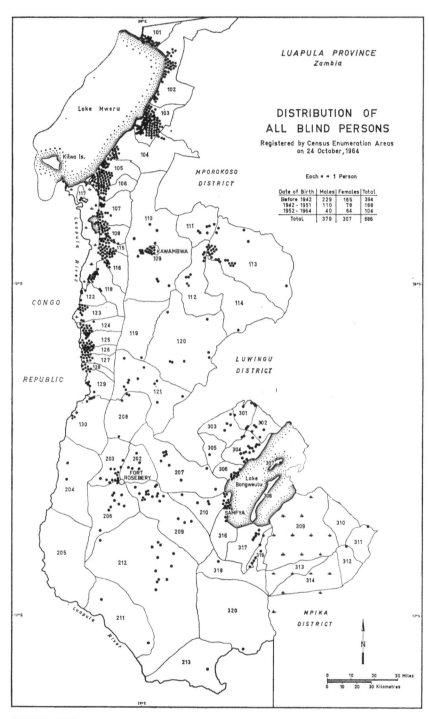

LUAPULA PROVINCE
Zambia

DISTRIBUTION OF
ALL BLIND PERSONS

Registered by Census Enumeration Areas
on 24 October, 1964

Each • = 1 Person

Date of Birth	Males	Females	Total
Before 1942	229	165	394
1942 - 1951	110	78	188
1952 - 1964	40	64	104
Total	379	307	686

FIGURE II.2

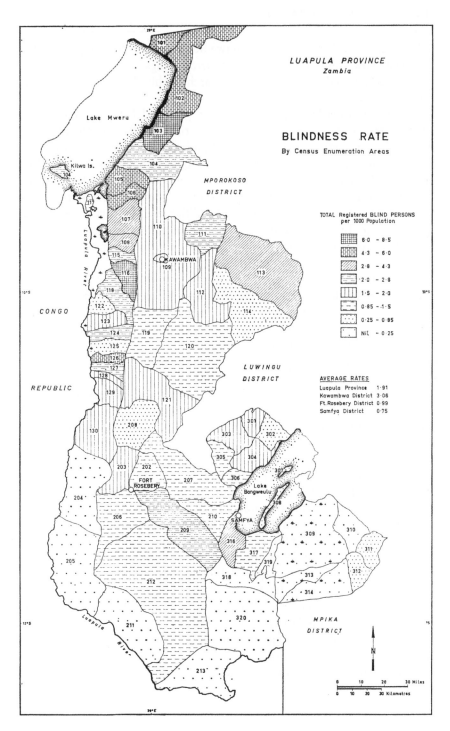

LUAPULA PROVINCE
Zambia

BLINDNESS RATE
By Census Enumeration Areas

TOTAL Registered BLIND PERSONS
per 1000 Population

	6·0 – 8·5
	4·3 – 6·0
	2·8 – 4·3
	2·0 – 2·8
	1·5 – 2·0
	0·85 – 1·5
	0·25 – 0·85
	Nil – 0·25

AVERAGE RATES
Luapula Province   1·91
Kawambwa District   3·06
Ft.Rosebery District 0·99
Samfya District   0·75

MPOROKOSO DISTRICT
LUWINGU DISTRICT
MPIKA DISTRICT

CONGO
REPUBLIC

Lake Mweru
Kilwa Is.
Luapula River
KAWAMBWA
FORT ROSEBERY
Lake Bangweulu
SAMFYA

0   10   20   30 Miles
0   10   20   30 Kilometres

FIGURE II.3

Mabumba (209) and Kasoma Bangweulu (316) are interesting since they may well show higher rates than their surroundings because of their proximity to urban medical centres. Here search for blind persons would have been far more thorough.

The average rates by district (see Fig. 11.3 key and Table 11.1) which, by virtue of greater numbers, are of greater statistical reliability than those of the small enumeration areas considered individually, show a most marked difference to the extent of a three times higher rate between Kawambwa district in the north and its two southern neighbours. In the latter districts the rates are not untoward even by developed countries' standards. (The all-England rate is 0·7/thousand.) The difference, however, lies in the ages affected. English blindness is mainly in the aged; throughout the Luapula onset is in the very young,[17] 62 per cent becoming blind before the age of seven and 76 per cent before thirteen.

The geographical analysis, which is here in part described, went on to look at three further aspects of the blindness problem.[18] Because measles was recorded as associated with onset of blindness in 52 per cent of all the 686 blind cases, these patients were separately mapped with mainly negative results. Surviving measles-blinded persons could not be shown to correlate with density of population. The factor of density might have been expected to affect diffusion of an epidemic.

Blindness amongst the different major cohorts of the census was also separately mapped. Here there is some slight cause for optimism. Although little spatial difference is recognisable, and today's generation are, in the main, going blind in much the same areas as their fathers, there is a marked overall improvement in blindness rates, especially amongst males. This may be because men are less conservative and more apt to seek treatment early. It also seems likely that improvement should be credited to the tireless efforts of the several Christian missions in the valley areas who, over two generations, have guided and advised the people and treated their sick. At places remote from this influence no improvement of rates is discernible.

Thirdly the study utilised specially collected subsidiary data on uni-ocular blindness amongst Luapula school children. The numerical breakdown by cause of uni-ocular handicap agreed closely with the registered causes of binocular blindness, but the rates by districts (8·3–10·5/thousand) for the children sampled indicate little local variation between geographical areas. Girls, 7·7/thousand, are less affected than boys, 10·7/thousand by uni-ocular blindness. In particular, boys apparently lose the sight of one eye more often than girls as a concomitant of smallpox (30 boys to 17 girls), measles (15 to 11) and chickenpox (6 to 1).

REFERENCES

1 DIXEY, F. (1945) 'The geomorphology of Northern Rhodesia'. *Trans. geol. Soc. S. Afr.*, **47**, 9–46.
 DIXEY, F. (1958) 'The geomorphology of Central and South Africa'. *Trans. Geol. Soc. S. Afr.*, Annex. 1958.

2 KING, B. C. (1957) 'The geomorphology of Africa'. *Science Progress*, **45**, 672–80.

3 DEBENHAM, F. (1949) *Study of an African Swamp*. Report of the Cambridge Expedition to the Bangweulu Swamps, Northern Rhodesia. H.M.S.O.

4 BRELSFORD, W. V. (1956) *The Tribes of Northern Rhodesia*. Lusaka.

5 MCGLASHAN, N. D. (1965) 'The Distribution of Blindness in the Luapula Province (Northern Rhodesia)'. Unpublished Univ. of London M.A. Thesis.

6 KAY, G. (1967) 'Maps of the distribution and density of African Population in Zambia'. University of Zambia Institute for Social Research Communication No. 2.

7 TRAPNELL, C. G. (1953) *The Soils, Vegetation, and Agriculture of North-Eastern Rhodesia*. London.

8 BURTON, SIR R. F. (1873) *The Lands of Cazembe, 1798*. Murray.

9 TILSLEY, G. E. (1929) *Dan Crawford of Central Africa.* Edinburgh.

10 LAMMOND, W. (1963) Personal Communication.

11 PHILLIPS, C. M. (1959) 'Blindness in the Kawambwa District'. Rhodes–Livingstone Institute Communication No. 15.

12 Laws of Northern Rhodesia, Cap. 207 (1961) *Notes for the Guidance of Authorised Officers* (Section 10–4).

13 COBB, B. (1962) 'Blindness and eye disease in the Luapula Province of Northern Rhodesia'. Unpublished Report to R.C.S.B.

14 AWDRY, P. (1963) 'Central report on the incidence and causes of blindness in the Luapula valley and adjacent areas in Northern Rhodesia'. Unpublished Report to R.C.S.B.

15 ADAMS, P. C. G. (1964) Personal Communication.

16 KAY, G. (1962) 'A population map (1 : 500,000) of the Luapula–Bangweulu Region of Northern Rhodesia'. Rhodes–Livingstone Institute Communication No. 26.

17 MCGLASHAN, N. D. (1969) 'Measles, malnutrition and blindness in Luapula Province, Zambia'. *Trop. geogr. Med.*, **21**, 157–62.

18 MCGLASHAN, N. D. (1966) 'Blindness in the Luapula Province'. *Cent. Afr. J. Med.*, **12**, 41–7, 68–73, 86–9.

12 The Decline of Malaria in Trinidad[*]

L. S. Fonaroff

Trinidad has long been recognised by the natural sciences as an 'island laboratory'. Its position just beyond the mouth of the Orinoco River makes it a 'continental island'. Its geological and biological affinities to the mainland are fairly well known and are continuing to be investigated by a wide variety of workers. Now after decades of work biogeographers can say, with reasonable certainty, that virtually all of the plant and animal life have their origins in one or another of the adjoining neotropical land areas.[1] This is a matter of record which in many instances is able to provide impressive detail. The only important intrusions to this Darwinian world, at least from a health point of view, are the appearance of Europeans and, subsequently, the malarial parasite.

The human population are a mixture of peoples of African and East Indian ancestry and are newcomers to the island. In the four and a half centuries since Columbus waded ashore at Moruga Bay, the indigenous peoples, occupiers of the island for millenia, have disappeared. The uniqueness of Trinidad's recent demographic history is too well known to repeat here. Suffice it to say that the various cultural groups presented diverse exposures to malaria, the disease they transported from the Old World in post-Columbian times.[2, 3, 4]

This paper demonstrates, in map form, the changing distributional character of malaria and offers tentative explanations for it.

RECENT INCIDENCE OF MALARIA

Much has been written in recent decades on Trinidad's public health problems. Pulmonary tuberculosis, for example, which had been a leading cause of death for more than a century, has been reduced to little more than a trace during the last fifteen years. Malaria, the other major cause of death and sickness, and the more serious of the two, experienced a more abrupt extinction in recent years despite its more devious natural history. Declining mortality statistics reflect the results of a multi-faceted approach to its eradication (see Table 12.1).[5, 6]

Malaria was not declared a notifiable disease by the local government

* Reprinted from *The West Indian Medical Journal*, (1968) **17**, 14–21.

TABLE 12.1

Malaria Mortality and Morbidity 1938-6

Year	Deaths/100,000	Notified Cases
1938	108·6	19,050
1939	96·8	(records unavailable)
1940	90·0	20,691
1941	97·7	15,835
1942	107·3	17,097
1943	113·8	18,196
1944	87·3	12,356
1945	76·8	9,455
1946	62·7	8,854
1947	37·2	6,115
1948	29·5	5,198
1949	24·7	4,827
1950	22·3	5,098
1951	21·3	5,641
1952	12·1	5,391
1953	10·9	5,050
1954	9·0	5,514
1955	6·1	1,540
1956	2·8	274
1957	0·8	640
1958	0·5	376
1959	–	96
1960	0·2	11
1961	0·1	2
1962	–	2

until 1956; nonetheless archival health records show numbers of people notified as having suffered from the disease for some years. With due regard to under-reporting, particularly morbidity, and other possible minor sources of error, malaria prophylaxis reduced known cases from 20,691 in 1940 to 11 in 1960. Deaths for the 1945–54 decade dropped from 425 to 63 for the respective years with the eradication effort in the control stage. Notified cases during the second half of the decade averaged little more than 5,000 per year.

MALARIA DISTRIBUTION

The maps illustrate not only the progress of a systematic eradication effort, but also provide a generalised spatial overview of malaria distribution on the island. This notion of 'spatial reality' as Kant once called it, in contrast to 'point reality', is of obvious epidemiological interest. It is believed that they would also be of interest to regional planners, medical geographers and ecologists, and historians. The item depicted on the maps is spleen enlargement, a fairly common index in malaria surveys. They represent occurrence as recorded in school children, presumably the most sensitive segment of the population. It should also be borne in mind that malaria is but one of more than fifty possible causes of spleen enlargement. Extensive ecological field studies, coupled with a thorough search of surviving health records and historical land-use documents, indicate that the depicted spleen values do in this instance represent malaria infection.

It should be recognised that any map fundamentally reflects the judgement of its maker with respect to the spatial distribution of the element or value being mapped. The judgement expended on these maps has included not only basic epidemiological considerations relating to the clinical and ecological character of malaria transmission but also the relevancies of the physical environment. The observations and data collection were made on the island in 1965. Data for the maps were 'smoothed out' from detailed yearly spleen surveys undertaken by the health ministry. It is believed that a network of more than seventy-five stations used in establishing the map-base has provided a reasonably true picture of the changing character of Trinidadian malaria.

Continuous spleen values over a broad geographical area are somewhat unusual. However, due to the high population density of the Caroni Plain, an unusually dense network of malaria recording stations developed. The mappability of these data thus appears to be not only justifiable, but provide an accuracy almost never approached on meteorological charts. They are, in essence, 'isosplenic' maps.

CHANGING SPLEEN RATE GEOGRAPHY

The spleen survey during the early 1940's carried out in conjunction with the Rockefeller Foundation programme revealed a surprising situation. More than half of the island was free of malaria. This is still very apparent in the earliest map (Fig. 12.2) for 1945. In comparing the 1913–14 spleen census by Lassalle,[6] an extremely contrasting situation appears. Although a considerable amount of malaria was to be found across the island in 1913, the thirty-year interval between surveys shows a highly significant shift in spleen rates from south to north. This early survey data, in the crude isolated form in which it is now available, has yet to be reduced to map form pending evaluation of data and techniques of collection. Nevertheless, several characteristics support the notion of the shift. The overall spleen rate for St. Patrick county on the south-western peninsula for 1914 was 57 per cent, but only 9 per cent in 1942. Health Ministry records likewise recorded a change in rates for St. David county on the north-east coast:

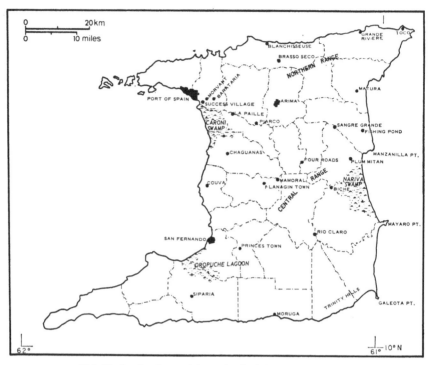

FIGURE 12.1 Trinidad: physiographic and administrative areas.

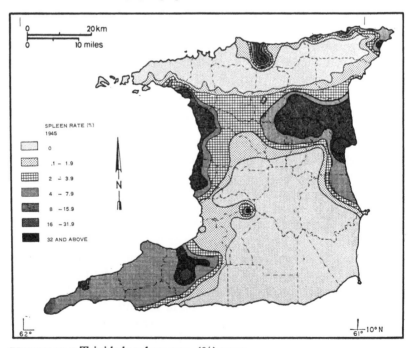

FIGURE 12.2 Trinidad: spleen rate (%) 1945.

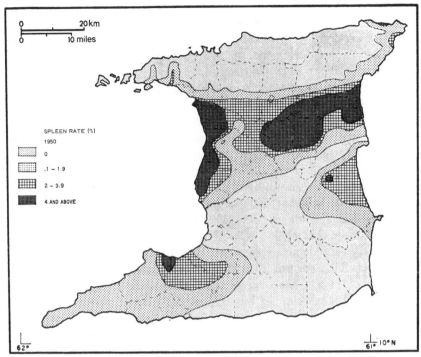

FIGURE 12.3 Trinidad: spleen rate (%) 1950.

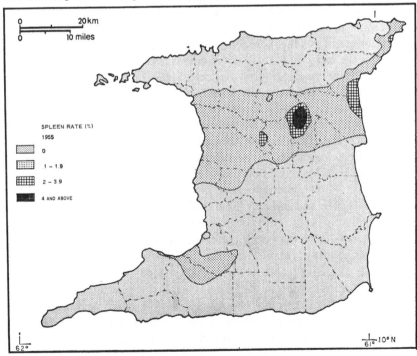

FIGURE 12.4 Trinidad: spleen rate (%) 1955.

17 per cent in 1913, 53 per cent in 1942. In fact the school rate for north-east coast schools during the 1940's occasionally exceeded 80 per cent. Virtually all stations show the reversal.

In more recent decades, the build-up of the cocoa industry in the central portions of the island precipitated the development of an endemic pocket of malaria due to the presence of *Anopheles bellator*, the bromeliad-breeding vector. In general, however, while several tentative partial explanations for the shift in spleen rates suggest themselves (such as population growth patterns, the 'premunition' factor, vector geography, etc.), the actual processes by which this has occurred are still not clear. The only obvious fact is that there, as in many malarial environments elsewhere, the disease frequently has a tendency to vary in intensity despite the early efforts expended on its control.

The 1945 map depicts spleen rate distributions a few years after the initial Rockefeller survey. This style of presentation also tends to reveal something of the vital characteristics of the disease. It suggests the presence of at least two overlapping but ecologically distinct vectors. As Major Senior-White discovered years ago, *Anopheles aquasalis* prefers brackish water for breeding and is found in virtually all coastal areas tested. Its capacity to keep transmission going was a function of an exceedingly high pupation and hatching rate. The comparatively high rate in the Port-of-Spain area is attributable to the unusual qualities of *aquasalis* and the topographic proximity and placement of the city itself. The vector has unusual strength as a flier with a range of three to five miles, possibly more. It also has the capacity to fly high enough to overcome Morvant Ridge, the only significant physiographic obstacle separating the Caroni Swamp (Laventille Section) from the south-eastern urban fringe (Fig. 12.1). Environmental lapse-rates appear to be critical in preventing the vector from becoming established in the high mountain areas of the Northern Range. The species does not normally occur above 650 feet elevation. Several climatological aspects of this problem are still under investigation.

With the possible exception of the Port-of-Spain area, all coastal areas with rates of 16 or higher appear to have been due to the presence of two vectors, *aquasalis* and *bellator*. The combined vectorial capacities of these two individually inefficient species is remarkable.

Anopheles bellator, the vector commonly found in cacao plantation areas, has a considerably complicated life history. It oviposits in epiphytic bromeliads exclusively and, as Pittendrigh demonstrated,[7] was responsible for malaria transmission in the plantations of the central highlands. High interior spleen rates are due to this species. The nearly stationary spleen values along the interior foothills of the Northern Range, and persisting through the 1955 map, appear to reflect the increasing daily round trip traffic and settlement pattern developing from Port-of-Spain and Arima and beyond, exposing travellers to both vectors. Several physiographic relationships of this vector are currently being developed, and the association of the vector with agricultural extension for citrus cultivation is under examination. Post-war agricultural development on the Caroni alluvial plain may be a factor in *bellator* malaria reduction.

The remaining outstanding anomaly on the 1945 map is the exceedingly high rates (32) in the north central mountain area. Ecological studies a decade or more ago have attempted to incriminate a third possible vector, *Anopheles homunculus*. Pittendrigh's researches on the ecotypic specialisation of *homunculus* and *bellator* have persuasively argued that in humid areas, *homunculus* is able to transmit malaria.[7, 8] This point of view is supported by circumstantial evidence and theoretical considerations of an evolutionary nature, although none of the specimens of homunculus dissected through the years has shown infection.

SUMMARY AND CONCLUSIONS

The Trinidad malaria eradication effort progressed through the years with reasonable speed and efficiency despite incomplete life history data of the vectors involved.

Figs. 12.2, 12.3 and 12.4 graphically show the declining dimension of the problem during its final stages. They also have the instructive potential of providing the investigator with a spatial appreciation of a problem not easily obtainable in statistical or tabular form. Used in combination with other types of maps they may suggest 'process' and indicate research angles not readily obvious to field workers.

ACKNOWLEDGEMENTS

Dr. R. M. F. Charles, director of Preventive Medical Services for the Government of Trinidad and Tobago, kindly provided field assistants, Health Ministry records and the use of a microbiological laboratory. The help of Messrs. Enal Ramgoolan and Hydar Ali of the Insect Vector Control Division is also gratefully acknowledged.

Field work was supported by U.S. Office of Naval Research, Contract NR–3656(83), Project NR 388–067, through the University of California, Berkeley.

REFERENCES

1 DARLINGTON, P. J. (1953) *Zoogeography: The Geographical Distribution of Animals*. New York.
2 BLUMBERG, B. S. (ed.) (1961) *Proceedings of the Conference on Genetic Polymorphisms and Geographic Variations in Disease*. New York.

3 JARCHO, S. (1964) *Bulletin of the History of Medicine*, **38**, 1–19.

4 MOTULSKY, A. G. (1960) *Hum. Biol.*, **32**, 28–62.

5 Government of Trinidad and Tobago (1963) *Annual Statistical Digest*. Government Printing Office, Port-of-Spain.

6 LASSALLE, C. F. (1916) *Trinidad Malaria Report*. Government Printing Office, Port-of-Spain.

7 PITTENDRIGH, C. S. (1950) *Evolution*, **2**, 43–63 and (1950) *Evolution*, **4**, 64–78.

8 DOWNS, W. G. and PITTENDRIGH, C. S. (1949) in Boyd's *Malariology*, pp. 736–48. Philadelphia and London.

13 Probability Maps of Leukaemia Mortalities in England and Wales

R. R. White

INTRODUCTION

This chapter may help to serve three purposes. The first is to focus on the peculiarly geographical problem of selecting an appropriate scale for the display of medical data in cartographic form. The second is to examine the utility of probability maps in measuring the appropriateness of a given scale. The third is to examine the spatial characteristics of the occurrence of leukaemia mortalities in England and Wales and to use these data to illustrate the first two aims.

MEDICAL GEOGRAPHY AND THE SCALE PROBLEM

It has been suggested that the principal task of the medical geographer is to determine where diseases are located and to relate such locational characteristics to relevant features of the physical and man-made environment.[1, 2] It is to the first of these tasks that this chapter is addressed. In particular this chapter is related to the problem of assessing the statistical significance of areally allocated data where the data are represented in choropleth form. Choropleth maps may be defined as being comprised of discrete areal units which exhaust a data surface without overlapping. The most common type of such maps is that based on administrative areas.

If a geographer accepts the responsibility of mapping the occurrence of morbidity and mortality he must also accept the statistical consequences of his decision to do so at a particular scale. To the geographer, 'scale' generally signifies 'scale of areal unit'. Of course, scale decisions are also involved in the choice of the time period over which the data are collected, or the taxonomic definition of a disease (e.g. can we speak of leukaemia as a single entity?). It is clear that the assessment of the significance of any observed pattern in an areal arrangement is dependent on the areal scale selected for representation. Harvey writes:

> It has been generally agreed (though not always observed in practice) that different processes become significant to our understanding of spatial patterns at different scales. For the most part, however, we have no measure of the

scale at which a particular process has most to contribute to the formation of a spatial pattern and our notions regarding the scale problem remain intuitively rather than empirically based.

and,

The analysis of the properties of spatial patterns at a variety of scales can also help to identify the scale at which a particular process is most effective.[3]

The probability map, as presented by M. Choynowski, may be one means of identifying some of the significant characteristics of data at any given scale.[4]

MEDICAL GEOGRAPHY AND PROBABILITY MAPS

Significance may be measured only in terms of some *a priori* expectation; thus an observation may be said to be significant to the observer if its occurrence surprises him.[5] Thus if one has no reason to anticipate different death rates between two areas, there should be some way of estimating the significance of the differences that do occur, i.e. do any of the observations for the individuals belonging to a set of areas differ significantly from the mean value for the set as a whole?

There are several reasons why one might anticipate that, for observations of any phenomenon over space, no significant difference would be observed. First, there might be no discernible reason why a given phenomenon should vary over space. For example, one would not expect the percentage members of a human population who were born on a Tuesday to vary significantly across a country. Second, even if significant differences might be anticipated for observations across a country, the demonstration of such significance may be highly dependent on the selection of the scale of the areal unit. For example, there is a positive correlation between mortalities from bronchitis and the amount of sulphur dioxide in the air, but this may be apparent only at a scale which differentiates urban areas from rural areas.[6, 7] Third, the magnitude of any 'real' difference between occurrences might be completely obscured by measurement errors.[8, 9]

Given any one of these conditions, or any combination of them, it is certainly possible that a group of observations based on areal data might appear to be random or insignificantly different from one another. Leukaemia, for instance, is a disease of which the aetiology is sufficiently obscure that it is not immediately clear at which scale it might be most appropriately mapped. In particular, the question as to whether leukaemia cases tend to cluster has not been answered.[10] Given (in the absence of evidence to the contrary) that leukaemia might be expected to occur randomly in space *and* that leukaemia mortalities are a rare occurrence for the population as a whole, it seems reasonable to test its observed occurrence against a random expectation based on the Poisson distribution. The Poisson probability distribution describes the likelihood of the occurrence of rare, random events in a continuum, given a mean expectation and a variance equal to that mean. A comparison of observed values with the values predicted by the Poisson generating function gives a measure of the likelihood of any given observation occurring by chance.[11, 12]

A simple numerical example of this procedure may be taken from a hypothetical case where the mean expectation (over a given period of time, such as one calendar year) for a given areal unit is three deaths. (For the purposes of this example it is assumed that all areal units have the same size of human population.) In one unit we observe only one death. How likely is it that we might observe so small a number of deaths, given the average expectation? In other words, how likely is it that we should observe one death *or* zero deaths? To answer this question we need to compute the combined individual probabilities of observing one death *and* zero deaths. This sum is called a cumulative probability value. If we observed ten deaths, we would ask how likely is it that we should observe so many deaths, i.e. ten deaths or more? Then we would have to compute the sum of the individual probabilities for all integers from ten to infinity. Of course, the individual probabilities would tend to zero very quickly. In this case, with a mean of three, the probability of observing even ten deaths is 0.0008.[13] Individual Poisson values are given by:

$$P(x) = e^{-a} \cdot \frac{a^x}{x!}$$

where a = mean incidence for all areas
  x = the observed value in a given cell
  e = the constant, 2.7183

For the cumulative probability we sum the probability of observing all the individual values from zero to x, as $x < a$, in this example. Where $x = 1$ and $a = 3$, the cumulative probability of observing x or less is $\cdot05$ (for $x = 0$) plus $\cdot15$ (for $x = 1$), which equals $\cdot20$.[14]

These cumulative probabilities may then be plotted for each areal unit to provide values for probability maps. Such maps have been made for the occurrence of mortalities from brain tumours in southern Poland[4] and for mortalities from stomach cancer in South Africa.[1] The following is a discussion of the problems involved in applying this technique to the mapping of leukaemia mortalities in England and Wales (Fig. 13.1–13.4).[15]

THE OCCURRENCE OF LEUKAEMIA MORTALITIES IN ENGLAND AND WALES

The crude death rate from all forms of leukaemia in England and Wales rose from 39 per million in 1950 to 60 per million in 1966. Although it is a rare disease and although the apparent rate has risen partly due to improved diagnostic facilities, it has nonetheless continued to occupy a prominent place in medical research.[16, 17]

Leukaemia may be described as a cancer of the blood cells. It is far more common in a chronic than in an acute form, and, like most diseases, it is most heavily concentrated in older age groups. The age distribution of leukaemia is bimodal with a minor peak among the under five-year-olds. Many studies have been concerned exclusively with childhood leukaemia.[18] Unfortunately, the data given in the Registrar-General's Annual Statistical Review after 1950 do not distinguish mortalities in different age groups at the county level.[19] Therefore, the following remarks refer to administrative counties without distinction between actual incidence among different age groups. 'Expectation' for the total mortalities in any county is calculated simply on the basis of total population. Thus it is necessary to allow for the effects of the unequal spatial allocation of different age groups after the cumulative Poisson probabilities have been calculated. In this respect this study is much cruder than those which use age-standardised mortality rates.[20, 21]

A brief review of the suggested contributory explanation of the distribution of leukaemia will precede an examination of the available data. A comparison of the data with the proposed aetiology of the disease should then indicate the more likely among the various possibilities, or at least point to the type of data that would be necessary to test a particular hypothesis.

Several writers have observed certain persistent regularities in both British and American[22] data. These regularities include:

(1) Higher urban than rural incidence.
(2) A high male–female sex ratio.
(3) A steady increase with socio-economic class.
(4) A tendency to marked regional variation in rates on a statewide basis in the United States and on a county basis in England and Wales and on a nation-wide basis in Western Europe.[17]
(5) A marked clustering on a micro-scale such as city block, school and family.[23]

Despite the recognition of these apparent regularities, there has been a remarkable lack of agreement in the medical profession on the probable causes of leukaemia. The higher urban than rural rate might suggest that this disease, like lung cancer, may be partly due to urban air pollution.

However, in England and Wales, on a county scale, the more urban counties have a lower rate than some rural counties. The income differential in incidence may be the easiest to explain, as all diseases may be regarded as being *competitive* for mortalities. Thus, in poorer families there is a much greater likelihood of children succumbing to environmentally induced illnesses such as pneumonia. Among older people, bronchitis and influenza are often fatal, thus reducing the number that would be susceptible to leukaemia.

The micro-level clustering raises a major unanswered question about the communicability of leukaemia. Communicability would support the idea of belts of leukaemia moving across a country as a low-level epidemic.[23] It would also suggest a reason for the persistence of high rates concentrated in neighbouring English counties which have little in common other than proximity. However, this sort of contention cannot be tested with the data available from the Registrar-General's Reports, as the locational interval of the county data is so great that it filters out any pattern that might be present at a micro-scale.

It is possible that hereditary factors may be important, though few cases have been documented to establish this. In the 1950's it was thought that background, or low-level, radiation might have a contributory effect, but investigations never supported the idea in Britain.[24] Finally, it has been suggested that insecticides, some preservatives in foods and various toxins may be partial causes. A summer peak in the onset of leukaemia has been related to the use of insecticides.[17]

Many factors are possible contributors to spatial irregularities in the distribution of leukaemia mortalities. The procedure of ecological correlation (i.e. the correlation of one of several related variables with the occurrence of leukaemia mortalities) appears to be of doubtful value as an inferential tool in this case as there is a wide variety of possible contributors or causes and there is the related problem of determining the most appropriate scale at which to record the mortalities and the variables involved. This last problem is usually supplanted by the yet more intractable one of being limited to the recordings at a given scale. A complex example of the temporal-spatial scale selection problem lies in the difficulty of actually comparing the rates of several competitive diseases through time and space when the temporal variability can be so great that, in comparison, any spatial variation is very small.

Regardless of the shortage of knowledge of the essential characteristics of the recorded data, Poisson probability maps were constructed for leukaemia mortalities in the administrative counties of England and Wales for each year from 1950 to 1966 inclusive. This was done in the belief that much of the variation of these data on the time-space scale of the 'county-year' might be ascribed to random variations. Thus the probability maps would appear as filtered data through which only the non-random record-ings would appear. The year unit was subsequently aggregated to periods of two years and four years, and the probabilities were recalculated on that time basis. An attempt at spatial aggregation was ruled out because of the difficulty of determining rules for aggregating irregular administrative areas. Temporal aggregation was chosen as the only alternative. The 95 per cent level was selected arbitrarily as the cut-off point for significant differences between observed and expected values. In this way attention was directed to ten or less counties on each annual map, rather than to the arrangement of raw incidence figures for all fifty-four county divisions. Table 13.1 shows the number of 'significant counties' for each map at different levels of temporal aggregation. Examples of the effect of compar-ing incidence with probability maps (Fig. 13.1 and 13.2) and of successively aggregating the time unit for observations are given (Fig. 13.3).

TABLE 13.1 *Mean Leukaemia Incidence and the Number of 'Significant Counties' at Different Levels of Temporal Aggregation*

Year	Mean incidence per million	Number of 'Significant Counties'		
		1 year	*2 years*	*4 years*
1950	39	8		
1951	27	5	14	
1952	29	7		26
1953	50	8	17	
1954	47	10		
1955	50	8	8	
1956	53	5		15
1957	51	5	7	
1958	53	9		
1959	59	7	12	
1960	54	5		20
1961	60	10	15	
1962	58	7		
1963	61	4	12	
1964	58	5		8
1965	59	6	7	
1966	60	6		

Source: calculations based on *Registrar-General's Annual Statistical Review for England and Wales, Part I*, from 1950 to 1966.

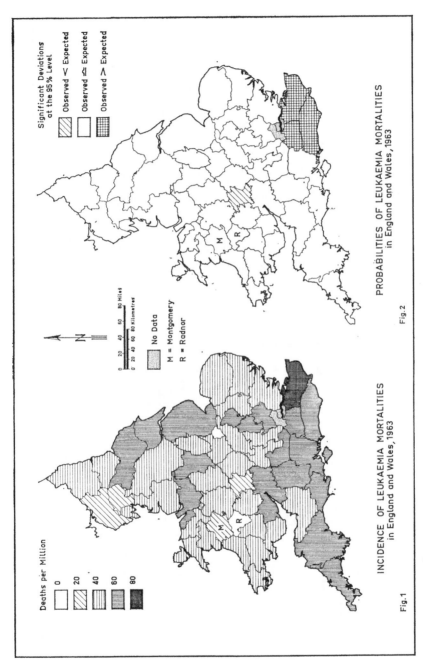

Deaths per Million

0
20
40
60
80

No Data

M = Montgomery
R = Radnor

0   20   40   60   80 Miles
0 20 40 60 80 Kilometres

N

Significant Deviations
at the 95% Level

Observed < Expected

Observed ≤ Expected

Observed > Expected

INCIDENCE OF LEUKAEMIA MORTALITIES
in England and Wales, 1963

Fig.1

PROBABILITIES OF LEUKAEMIA MORTALITIES
in England and Wales, 1963

Fig.2

FIGURES 13.1 and 13.2

G

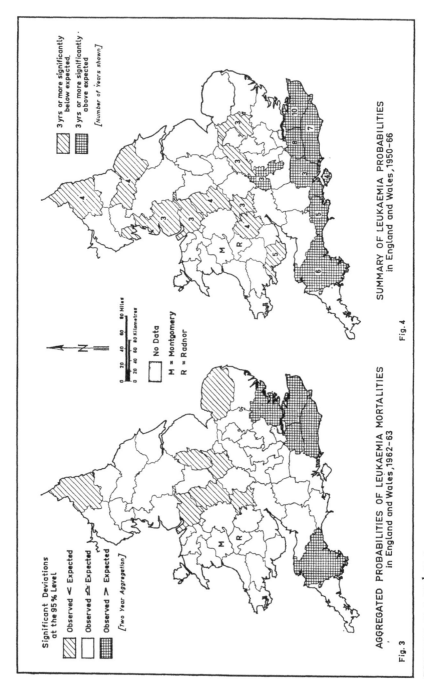

Significant Deviations
at the 95% Level

⧄ Observed < Expected

☐ Observed ⋜ Expected

▦ Observed > Expected

[Two Year Aggregation]

No Data

M = Montgomery

R = Radnor

0 20 40 60 80 Miles
0 20 40 60 80 Kilometres

3 yrs or more significantly
below expected.

3 yrs or more significantly
above expected

[Number of Years shown]

AGGREGATED PROBABILITIES OF LEUKAEMIA MORTALITIES
in England and Wales, 1962–63

Fig. 3

SUMMARY OF LEUKAEMIA PROBABILITIES
in England and Wales, 1950–66

Fig. 4

FIGURES 13.3 and 13.4

The importance of weighting the observations of mortality rates by the size of the human populations is illustrated by comparing Fig. 13.1 with Fig. 13.2. In 1963 Radnor had no leukaemia mortalities and therefore had an incidence of zero per million (Fig. 13.1); Montgomery had one death – an incidence of 23 per million – yet both of these fail to register as significant deviations from a random expectation (Fig. 13.2). It is true that the probability map for 1963 shows the fewest significant counties of any year (Table 13.1), but all the maps show substantially the same regularity. The smaller counties are rarely significantly different from the mean, while those counties that are significant on the probability maps show a more coherent 'pattern' than they do on any of the incidence maps.

Figs. 13.3 and 13.4 are examples of progressive temporal aggregation. Fig. 13.4 simply shows those counties that displayed a non-random deviation from the national mean for any three of the seventeen years from 1950 to 1966 inclusive. (The number of years for which they were above or below the mean is shown for each county.) A few counties were significantly above expectation one year and significantly below another; but none of these 'mixed' counties had three or more either above or below the line and therefore do not appear on the map.

The maps show a higher than expected value for several counties in the south-east and along the south coast. Areas of low incidence include Glamorgan and some of the industrial counties of the midlands and northern England. It may be noted that such a distribution is incompatible with any suggestion that leukaemia may be positively correlated, at this scale, with air pollution. Furthermore, it bears little similarity to the county distribution of old people or of higher income groups.

Temporal aggregation was attempted for two reasons. First, it might be thought that the small counties had little chance to exhibit a significant deviation from the mean as their populations were too small. One way to increase their populations at risk is to extend the time period.[1] Second, it has been suggested that the development of acute leukaemia operates on a 300-day cycle.[17] This, coupled with a summer maximum onset of the disease, might give an irregularity of incidence from one calendar year to another. An extension of the basic time period might therefore show a more interpretable pattern of significant occurrences. Table 13.1 does suggest that there is some increase in the number of significant deviations from a random expectation with increase in the size of the time unit, although it would be difficult to establish that the observed spatial arrangement becomes more interpretable as this aggregation proceeds.

Other aspects of the time scale that present difficulties in terms of aggregation are the observed trend in the mean since there has been an irregular increase in the incidence of leukaemia in this seventeen-year period (Table 13.1). More detailed data are also needed concerning the date of the onset of the disease for each individual. This latter feature would give inaccurate information on the nature of cycles of incidence of the disease where these are out of phase with the divisions of the calandar year. The problem of temporal trend makes comparisons from year to year very difficult, and a histogram showing the number of cases per year for any county would not be strictly correct as each year would not necessarily be comparable with the next.

A more complex question is raised by the observed increase in incidence, because this increase might represent different stages in the development of the disease. For example, two quite different spatial processes might be hypothesised:

(1) One might anticipate an initial irregularity (with a random or non-random spatial arrangement), followed by a later stage tending to a more uniform incidence, or

(2) the initial stage might be followed by a more 'mature' spatial arrangement which would follow some uneven distributional characteristic of the human population such as susceptibility to competitive disease. Unless an explanation for the major temporal variations can be found, then one cannot properly say whether one year may be considered as comparable (in terms of the stage of the hypothesised process) to another.

Despite these difficulties, the problems of establishing the statistical characteristics of any temporal scale are small compared with similar problems which arise from measuring *spatial* properties at different scales. It has been stated already that several hypotheses related to the spatial occurrence of leukaemia cannot be tested at the county scale of the data published for England and Wales. These are the hypotheses that are concerned with micro-clusters which suggest that leukaemia may be a communicable disease. Similarly, it would be meaningless to describe the spatial *spread* of a disease as rare as leukaemia at the county scale (although this was done by MacMahon[23] for states in the United States). The large size of the administrative county and the great variety of aetiological characteristics contained by each county may well be so great as to nullify the interpretive utility of even the probability based maps. The county boroughs, which represent wholly urban populations above a certain size, are far more homogeneous in aetiological characteristics and in leukaemia incidence. In 1950, of the eighty-three recorded boroughs only four were significantly non-random at the 95 per cent level; in 1960, only two of the eighty-three were significant. This confirms Hewitt's finding, based on earlier data (1950–2), that 'so slight are the differences between the county boroughs that regional variation can be shown to arise chiefly from variations in rates outside the main towns'.[16]

Linked with this situation is the third aspect of scale selection, namely the classification of the phenomena themselves. Leukaemia mortalities may be ascribed to one of three main types of virus – myeloid, lymphoid and monocytic. These types may well have different causes, and they certainly have different patterns of age-specificity. Also it would be valuable for leukaemia mortalities to be broken down by age rather than just by sex. If deaths are published simply for all types for all ages together, then it is impossible to take the analysis to a finer level.

Urban and rural population groups have been shown in other data to have significantly different incidences, but the Registrar-General discontinued the publication of the urban district and municipal borough versus rural district breakdown of the administrative counties after 1950. As many urban people do not live in county boroughs, it is not possible, using currently published data, to compare the county boroughs with administrative counties as representative of the urban and rural incidences respectively.

The problem of age-specificity can be circumvented partly by taking the national age breakdown of leukaemia incidence and the age structure of each county and then deriving an age-based mortality expectation for each county for each year. However, in addition to the computational labour involved by calculating separate expectations for fifty-four counties for seventeen years and six age groups, the results will give only a better expectation and will not solve the difficulty presented by the coarseness of the leukaemia data themselves. An example of this procedure is set out in

TABLE 13.2 *Age-standardised Leukaemia Mortality Expectations*

	Durham	Lancashire	Leicester	Sussex	Yorkshire West
Age Structure:					
% Age Groups –					
0–	9·0	9·0	9·0	7·1	8·9
5–	16·0	14·9	15·0	13·1	15·2
15–	26·2	26·0	27·5	21·8	26·3
35–	26·2	25·6	26·2	23·8	26·0
55–	11·6	12·4	11·1	14·0	12·0
65+	11·0	12·1	11·2	20·2	11·6
Leukaemia:					
Observed	57	140	22	63	91
Initial					
Expectation	58	140	26	51	104
Age-standardised					
Expectation	44	120	21	63	82

Sources: British Government, *Sample Census, 1966, England and Wales, County Reports 1–59* (London: H.M.S.O. General Register Office, 1967), British Government, *Registrar-General's Annual Statistical Review for England and Wales, Part I, 1966* (London: H.M.S.O.).

Table 13.2 for five counties of varied age structure for 1966, when population figures may be fairly accurately taken from the 1966 Sample Census.[25] The age-standardised expectation gives a much closer fit to the observed mortalities for Sussex and for Leicester, a slightly closer fit for the West Riding of Yorkshire, but a much worse fit for Lancashire and Durham. It might appear that simply improving the basis for expectation does not necessarily provide a better explanation, though five examples are an inadequate basis for this kind of inference.

All that may be induced at this level of expectation is that the probability maps do appear to confirm that there are persistent significant differences between the *per capita* rates of leukaemia deaths in different administrative counties of England and Wales. These differences cannot be related *at this scale* to urban-rural differences, income distribution, age distribution or air pollution alone. Almost certainly the only useful approach to the problem is multivariate, with data that are available on a large enough scale to discriminate between significant differences and those that can be ascribed to random variations.

REFERENCES

1 MCGLASHAN, N. D. (1966) 'The medical geographer's work'. *Int. Pathology*, **7**, 81–3.

2 MCGLASHAN, N. D. (1969) 'The nature of medical geography'. *Pacific Viewpoint*, **10**, 60–4.

3 HARVEY, D. W. (1968) 'Pattern, process and the scale problem in geographical research'. *Trans. Pap. Inst. Br. Geogr.*, **45**, 71–2.

4 CHOYNOWSKI, M. (1959) 'Maps based on probabilities'. *J. Am. Statist. Ass.*, **54**, 385–8.

5 PARRATT, L. G. (1961) *Probability and Experimental Errors in Science*. New York: John Wiley.

6 ASHLEY, D. J. B. (1969) 'Environmental factors in the aetiology of lung cancer and bronchitis'. *Br. J. prev. soc. Med.*, **23**, 258–62.

7 GARDNER, M. J. and WALLER, R. E. (1970) 'Environmental factors in the aetiology of lung cancer and bronchitis'. *Brit. J. prev. soc. Med.*, **24**, 58–60.

8 MCGLASHAN, N. D. (1969) 'The African lymphoma in Central Africa'. *Int. J. Cancer*, **4**, 113–20.

9 WOOD, E. (1960) 'A survey of leukaemia in Cornwall 1948–59'. *Brit. med. J.*, ii, 1760–4.

10 MERRINGTON, M. and SPICER, C. C. (1969) 'Acute leukaemia in New England: an investigation into the clustering of cases in time and space'. *Br. J. prev. soc. Med.*, **23**, 124–7.

11 MORONEY, M. J. (1962) *Facts from Figures*, pp. 97–107. London.

12 HOGG, R. V. and CRAIG, M. T. (1965) *Introduction to Mathematical Statistics*. 3rd ed, pp. 86–98. Toronto.

13 WHITE, R. R. (1970) 'Notes on some inferential problems on choropleth maps'. Mimeographed paper. Northwestern University.

14 General Electric Company (1962) *Tables of the Individual and Cumulative Terms of the Poisson Distribution*. Princeton University Press.

15 WHITE, R. R. (1970) 'Geographic Information and the Interpretation of Choropleth Maps'. Unpublished Ph.D. thesis, Department of Geography, University of Bristol.

16 HEWITT, D. (1955) 'Some features of leukaemia mortality'. *Br. J. prev. soc. Med.*, **9**, 81–8.

17 KNOX, G. (1964) 'Epidemiology of childhood leukaemia in Northumberland and Durham'. *Br. J. prev. soc. Med.*, **18**, 17–24.

18 EDERER, F., MANTEL, N. and MYERS, N. H. (1964) 'A statistical problem in time and space: do leukaemia cases come in clusters?' *Biometrics*, **20**, 626–38.

19 British Government, *The Registrar-General's Annual Statistical Review for England and Wales, Parts I, II and III*. London: H.M.S.O. Published annually.

20 ARMSTRONG, R. W. (1969) 'Standardized class intervals and rate computation in statistical maps of mortality'. *Ann. Ass. Am. Geogr.*, **59**, 382–90.

21 BRADFORD-HILL, A. (1966) *Principles of Medical Statistics*, pp. 201–19. 8th revised edition. London.

22 MCCULLOGH, J. J., ROSSOW, G. and HILLER, R. (1967) 'Leukaemia deaths in Minnesota 1950–64'. *Publ. Hlth Rep., Wash.*, **82**, 946–56.

23 MacMAHON, B. (1957) 'Geographic variations in leukaemia mortality in the United States'. *Publ. Hlth Rep., Wash.*, **72**, 39–46.

24 COURT-BROWN, W. M., *et al.* (1960) 'Geographical variations in leukaemia mortality in relation to background radiation and other factors'. *Br. med. J.*, 1753–9.

25 British Government, *Sample Census 1966, England and Wales. County Reports 1–59.* London: H.M.S.O.

14 European Male Stomach Cancer in South Africa: A Cartographic Appraisal

N. D. McGlashan

INTRODUCTION

Methods of mapping morbidity or mortality rates as commonly employed depend upon a calculation, within each chosen unit of area, which utilises the number of cases of disease or death in the period divided by the number of people at risk to that particular medical condition. A slight degree of sophistication can be added by dividing both the cases and the population into some form of breakdown on a basis of age and/or sex.

Such methods suffer from certain inevitable geographic disadvantages. In a country with wide variations of population density, the population totals at risk in the various subdivisions will also vary widely. Consequent variations in reliability will occur when areas are visually compared on the map. In general the meaningfulness will increase with increasing size of the population-at-risk. Such an increase can be achieved by two distinct means: aggregation by time or by area.

To increase the span of time covered by the map has certain attractions but also some disadvantages. The annual rate of disease (or death) in an area is calculated on a far surer basis with a larger effective population-at-risk if a period of several years is considered, and a longer time span also reduces the effects of random temporal fluctuations of numbers. On the other hand, the cost of registering cases increases with time, the analysis of the medical results is delayed and secular change of rates during the period of study is disguised by aggregating time.

To increase the size of geographical subdivisions by accumulating groups of initial local registration areas is also commonly used as a means of increasing the population-at-risk. This has the disadvantage of introducing subjective judgement into the grouping to be employed and also tends to mask the original geographical specificity of the data; that is, the register of case information may contain more accurate locational information than is utilised in a larger unit of area.

This paper is concerned with a specific example facing these problems.

THE MEDICAL DATA

Registration of the medically certified causes of death of the whole of the white population of the Republic of South Africa is recorded on computer cards by the Bureau of Census and Statistics in Pretoria. (The non-white population is excluded because certification of causes of death is not yet practicable or legally obligatory. Records are therefore unavailable.)

The registered deaths of males from Carcinoma of the Stomach (International Classification of Disease No. 151) were extracted for this analysis for a retrospective period totalling six years between 1949 and 1958 for which data were available. Death certificates in South Africa are recorded by Magisterial District of residence (with certain rules for the transference of deaths in institutions to be recorded under place of normal domicile). The Magisterial Districts have, in general, small population totals, but they are grouped into Economic Districts (E.D.) for government planning purposes. The forty-five economic districts have the advantage of being official units based on some degree of geographical homogeneity.

MAPPING THE DATA

Stomach cancer mortality is drawn up in Table 14.1 with economic divisions (column 1) arranged in order of increasing standardised mortality rates (s.m.r.)* (column 4) calculated from the observed cases recorded in the division (column 2) and the numbers expected. The expected figure is based upon a comparison with the national average rate adjusted for local population structure within age groups.

Applying the standard deviation (s.d. = 29·82) to the s.m.r. figures, one may class the economic divisions as away from the mean at ±1 s.d. and ±2 s.d. levels.[1] The position of the negative value of −3 s.d. is also indicated in the table.

Assuming, for the moment, that the s.m.r. values are normally distributed, the ±2 s.d. class interval which excludes 4·56 per cent of the normal curve would, in this case, give a significance value $p > 95$ per cent to only three E.D.'s, all of which lie in the 'low' tail. This system of categories is illustrated in Fig. 14.1, where the shadings represent the objective classes resulting from s.d. spacings.

It may be objected, however that this spread of s.m.r. values is not truly normal and, in fact, exhibits marked negative skewness. Thus a measure of significance based upon deviations from the mean is not entirely satisfactory. Furthermore the s.d.-based class interval treats all s.m.r. values as having equal reliability regardless of the actual number of observed cases used in the calculation.

An alternative method of classing the s.m.r. values for cartographic

*See appendix p. 194 for method of calculation.

TABLE 14.1 *Stomach Cancer Mortality*

1 Econ. Divn.	2 Obs.	3 Exp.	4 s.m.r.	5 S.D. 29·82	6 Scatter Class Limits	7 Poisson Signif.
				10·5(−3)		
					15	
32	1	5·4	18·4	*		**
35	1	5·4	18·5	*		**
31	8	20·2	39·6	*		**
				40·4(−2)		
36	9	22·0	40·9			**
63	3	6·2	48·1			
					50	
					60	
47	32	50·4	63·6			**
30	124	180·4	68·8			*
				70·2 (− 1)		
34	34	47·2	72·2			
51	21	29·0	72·5			
50	33	42·7	77·2			
					80	
18	52	64·0	81·2			
6	14	16·7	84·1			
16	39	46·3	84·2			
19	49	58·1	84·3			
9	21	24·4	86·1			
13	10	11·3	88·3			
1	241	267·1	90·2			
37	25	27·6	90·5			
					95	
2	80	82·4	97·1			
12	37	37·7	98·3			
40	320	320·6	99·8			(Close to
10	13	13·0	100·3			National
7	25	24·8	100·9			Norm)
66	30	29·5	101·6			
48	42	41·2	101·9			
61	19	18·4	103·3			
43	128	123·2	103·9			
					105	
64	84	79·6	105·6			
8	92	86·9	105·9			
44 ·	27	25·2	107·2			
62	80	73·1	109·4			
41	158	141·1	111·9			
17	43	38·4	112·1			
49	43	37·4	114·9			
3	35	30·1	116·2			
					120	
60	20	16·5	121·0			
5	43	35·0	122·7			
4	71	57·4	123·7			
67	29	23·3	124·4			
46	72	55·9	128·7			*
11	67	51·7	129·5			*
				129·8 (+ 1)		
					130	
42	82	61·5	133·4			*
45	121	90·6	133·5			**
15	11	7·9	138·9			
14	34	22·9	148·3			*
					150	
				159·6 (+ 2)		
TOTAL	2523	2523	100			

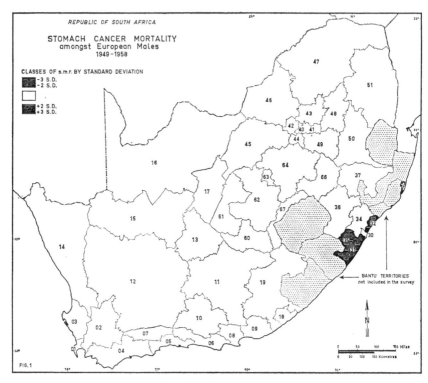

FIGURE I4.I

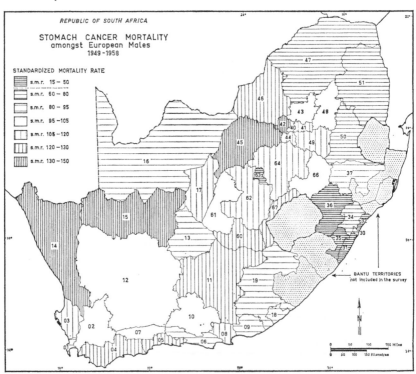

FIGURE I4.2

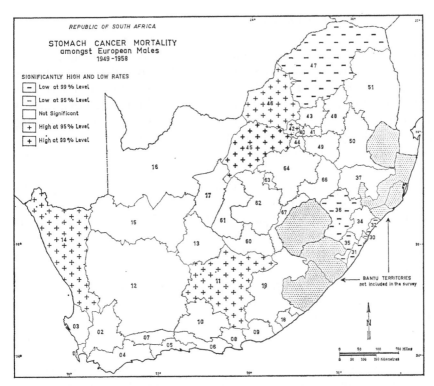

FIGURE 14.3

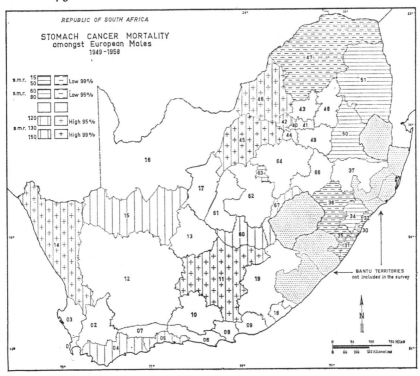

FIGURE 14.4

display is the more subjective method of using a scatter diagram to detect discontinuities for use as class boundaries, preferably with some semblance of symmetry about the mean. Seven resulting classes are shown in column 6 of Table 14.1 and also on Fig. 14.2, where it will be noted that there is a complete absence of s.m.r.'s falling in the s.m.r. 50–60 class. The shading scheme is designed to emphasise the high and the low ends of the scale since s.m.r.'s around the national norm (s.m.r. 80–120) are of little medical interest.

The diversity of sheer numbers arising from variations in size of populations-at-risk in different E.D.'s (mentioned above) can be exemplified from Natal with 124 cases in E.D.30 (Durban) but only a single case in E.D.32 (Natal North Coast). The E.D. with the largest number of cases, E.D.40 (Johannesburg) with 320 cancer deaths, happens to lie close to the national norm.

To allow for this, column 7 of Table 14.1 and Fig. 14.3 show statistical significance as derived from Poisson calculations. These assess in which E.D.'s the number of cases observed could have been expected to have been generated by random causes. In this case the rare event, a white male death from stomach cancer, is suitably approximated by the Poisson distribution. Importance is therefore attributed to those E.D.'s whose observed case numbers diverge from the expected numbers in either higher or lower direction at the 95 per cent[*] or 99 per cent[**] levels. Significance only arises in the top two and bottom two classes of Fig. 14.2, but exceedingly high or low *rates* by no means always warrant statistical significance.

In Fig. 14.4 Poisson significance and the arbitrary classes of Fig. 14.2 are combined and a further point is emphasised. Not only have the Natal E.D.'s (E.D. 30–7) significantly low rates but they also corroborate each other geographically by spatial contiguity.

In this actual example there is a considerable degree of differences between 'significance' assessed by the two methods: standard deviation from the mean in Fig. 14.1 and the Poisson formula in Fig. 14.3. The selection of a method of choice in a specific problem would depend upon firstly the normality of the s.m.r. curve and, secondly, the consistency and comparability of the sizes of numbers in the observed and expected columns. In general the standard deviation as a measure of significance is more likely to suit common complaints and diseases of high incidence.

DISCUSSION

The purpose here has been primarily to illustrate and discuss methods of preparation of maps meaningful to both physicians and geographers, but the distribution pattern of this cause of death is so striking that two comments may be offered.

Since U.K.-born migrants to South Africa retain the lower stomach cancer mortality rates experienced in Britain,[2] a high proportion of these migrants in Natal might contribute to the lower total rates illustrated. Unfortunately no division of cancer mortality by place of birth has been published and government records of immigrants cease after their first placing in employment. Many first settle on the Witwatersrand, and although they

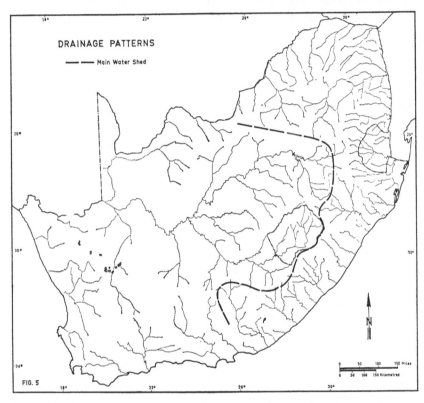

DRAINAGE PATTERNS

— · — Main Water Shed

FIG. 5

FIGURE 14.5

may later move to Natal, this has not been substantiated. The question of heredity or environment would then indeed be an interesting one.

An alternative aetiology is suggested by comparison of Fig. 14.3 with Fig. 14.5. Areas E.D. 47, 50 and 51 have one major geographical similarity to the other low incidence areas already mentioned which all lie in Natal, and which distinguish them from the high incidence areas elsewhere. The low incidence areas lie on the eastern side of the major continental water divide and are drained by the Crocodile, Tugela, Sundays and Fish River systems eastwards into the Indian Ocean. The rest of South Africa drains by the Vaal–Orange system from Lesotho westwards to the Atlantic and from the Cape mountains southwards by a number of lesser streams. As water supply has been put forward as having a possible link with stomach cancer causation this may well be important as a geographical clue[3]. If the clue is domestic water, it might be useful to survey the types of agricultural fertilisers used and to gauge the amounts of trace contamination which can be detected in various river systems. Areas E.D. 45, 46 and 42 in the western Transvaal certainly use large quantities of nitrate and phosphate fertiliser for wheat production. Alternatively mining industry effluents might allow pollutants to reach the water supplies of the western Transvaal and northern Cape Province.

APPENDIX

Expected Mortality

The purpose is to calculate, for each geographical subdivision, the 'expected' number of deaths appropriate to that area on the basis of two assumptions: firstly that the national average rate of death is evenly spread over all areas, and secondly that allowance is made for local variations of populations within different age and sex groups. For instance a retirement suburb may be expected to show far lower absolute death rates for complications of childbirth than a developing suburb with many young married couples.

International practice is generally to use ten divisions of each sex by age, In this paper six male (only) divisions are used, namely 0–34, 35–44, 55–64. 65–74 and 75+. This choice of age groups is controlled by the age groupings chiefly affected by the disease under study.

During the period 1949–57, the proportions of the total white male population of South Africa were calculated to give six divisions of a standard population million. (These proportions were 572,937; 142,341; 115,083; 88,221; 55,840; 25,578.) The total recorded (or 'observed') male deaths from stomach cancer in the period were 2,523 in all age groups and their division into the six age groups was also known. These figures can then be used to calculate national mortality rates for each age group severally and the rate for all age groups is the additive total of these six rates.

The rates for each age group can be applied to the population within each age group within each geographical area and totalled to give the expected number of deaths for that area.

The Standardised Mortality Rate[4]

To allow direct comparison to be made between areas with populations diverse both in total number and in age structure, the standardised mortality rate (s.m.r.) is calculated. The total number of deaths actually observed is divided by the calculated figure 'expected' (from above) and multiplied by one hundred for each geographical subdivision.

The national norm will therefore be 100 and s.m.r.'s above 100 will occur where more than the expected deaths are recorded. An s.m.r. below 100 indicates that recorded, or observed, deaths are fewer than have been postulated as 'expected'.

ACKNOWLEDGEMENTS

I wish to thank the South African Bureau of Census and Statistics in Pretoria for the basic data and to acknowledge my personal debt to the late Dr. A. G. Oettlé of the South African Institute for Medical Research for encouraging my interest in the problem here discussed.

REFERENCES

1 ARMSTRONG, R. W. (1969) *Ann. Assoc. Am. Geogr.*, **59**, 382.
2 DEAN, G. (1965) *S. Afr. med. J.*, **39**, suppl. 1–20.
3 OETTLÉ, A. G. (1964) *J. natn. Cancer Inst.*, **33**, 383.
4 HILL, A. B. (1966) *Principles of Medical Statistics*, Ch. 17, 8th Ed. London.

Part IV
Associative Occurrence

15 Geographical Evidence on Medical Hypotheses*

N. D. McGlashan

INTRODUCTION

The geographer's contribution to medical knowledge can properly be expected to lie chiefly in the field of environmental studies and the relation of disease distribution to other geographic variables.

In this context it is helpful to distinguish between physical environmental factors and the man-made or cultural environment. Studies in the first category include Ordman's[1] work on the relationship of respiratory allergy to altitude and Webb's[2] work on geochemical threshold levels for waterborne parasites.

Social environment may be difficult to assess quantitatively but must include those customs of man which are, so to speak, self-inflicted. Drinking and smoking habits would be classified here as would marathon-running or circumcision. Interest in this group of socio-environmental variables is justified because here the opportunity for variation is at its widest and also at its least well documented. It may be in this field that Africa has most to offer.

Geographers have, to date, been only uncommonly connected with work of this type and there is an urgent need for quantitative studies geared to permit correlational values to be calculated. The developing world will not postpone its social advance until geographers are ready with a metric technique.

This paper therefore presents a method that has been tried in certain African territories. Its biggest failing is the inevitably broad sweep of generalisation needed to utilise data of the kind recorded. In areas with more sophisticated medical and social records there must be correspondingly more opportunity to put similar or more refined statistical techniques into use.

The interest to a geographer lies chiefly in the comparison analytically of pairs of distribution maps. Whilst Figs. 15.1 and 15.2 and Figs. 15.3 and 15.4 give an immediate appearance of some similarity it is required to know,

* Reprinted from *Tropical and Geographical Medicine*, (1967) **19**, 333–43.

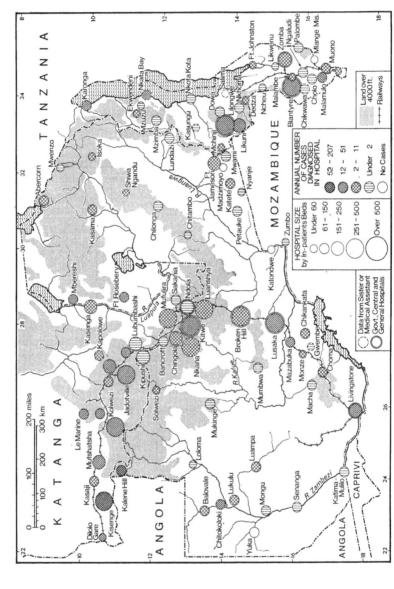

FIGURE 15.1 Central Africa: Diabetes Mellitus.

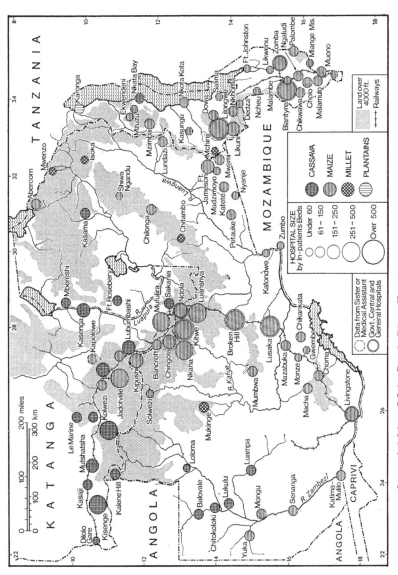

FIGURE 15.2 Central Africa: Main Staple Plant Food.

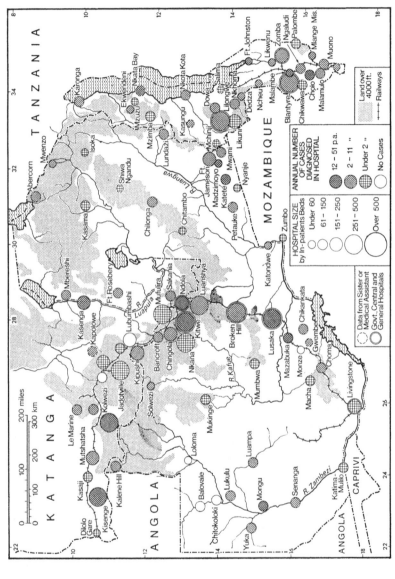

FIGURE 15.3 Central Africa: Cancer of the Uterine Cervix.

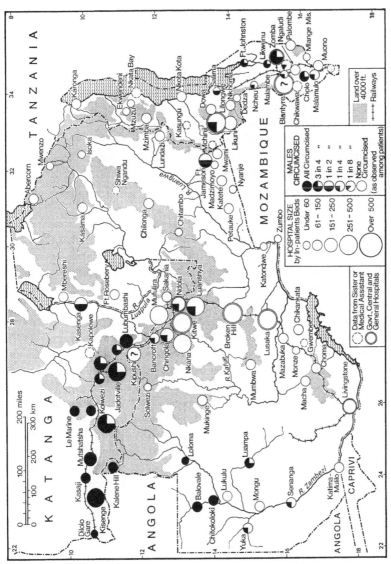

FIGURE 15.4 Central Africa: Circumcision of Males.

for example, which of these pairs shows the higher degree of spatial association. Such a method would also allow pairs with a low similarity rating
to be discarded.

LIMITATIONS OF ACCURACY

Whilst the data presented here are genuine in the sense that they were
honestly reported and compiled, the reader is advised that 'impressions of
disease cases usually diagnosed in an average year' from each hospital left
considerable scope for medical inaccuracy and that medical 'findings' (or
implications) should be looked upon with caution.

A geographical survey of fifty-five diseases and some twenty 'geogens'*
amongst African patients in territories of central Africa was intended
firstly to give an outline of their geographical distribution and secondly to
allow a quantitative assessment of the degree of spatial similarity between
pairs of factors.

Since information on each item reported, whether disease cases or indigenous custom, was necessarily subject to wide margins of error it must be
stressed that these maps can merely purport to show the broadest outlines
of spatial variation. If interest, be it medical or geographical, emerged it
should be thought of as suggestive of further lines of inquiry and very
definitely not as an end in itself.

Disease maps (Figs. 15.1 and 15.3) were compiled to show the numbers
of cases of each condition reported by each hospital in an average year. No
attempt was made to relate the cases to population served. The question of
population served by any hospital is difficult anywhere in the world since
it usually depends upon each patient's free choice. In Africa his first choice
is between western and indigenous medicine. Not only is the total population of some areas unknown, but even if population figures existed, it could
not even be guessed what number would actually seek hospital treatment
when sick. For reasons therefore of unknowns, 'incidence' (calculated as
cases divided by population) is not here quoted, and no allowance for
population served is included in this paper. It was recognised that this
omission was statistically serious. The only corrective included here lay in
the caution with which the statistical evidence was finally to be weighed.
The size of each hospital by in-patient capacity (itself an inadequate
measure) is shown by proportional circular symbols on these maps. Beds
earmarked for maternity, mental, tuberculosis or other chronic or long-
stay patients are excluded so that the remaining number, which is that
shown, can be regarded as general use beds.

---

* A geogen is defined by May[3] as 'an environmental factor known or believed to be
correlated with a disease or its pathogens'.

EXAMPLE A: DIABETES MELLITUS AND CASSAVA EATING

Existing medical opinion[4] on the causative factors of diabetes suggests either that hereditary factors are important or that dietary habits, in some way not yet understood, bring about an impaired function of the patient's ability to absorb and utilise glucose. Against this background a check of the correlation between hospitals reporting diabetes and those reporting the eating of particular foods might be informative. In this example cassava as the local staple plant food was considered.

The category sizes in the tables are related to the calendar; over four cases per week (208 per year), over one case per week (52 per year), over one case per month (12 per year), over one case in six months (2 per year), under one case in six months and no cases diagnosed. No hospital in the area reports over four cases of diabetes mellitus per week and (in Tables 15.3 and 15.4) no hospital reports over one case per week of cancer of the uterine cervix.

In Table 15.1 the observations from all the eighty-four hospitals are set out and this table is rewritten as Table 15.2 so that all 'expected' cell values are greater than 5. The numbers within the cells in all tables represent

TABLE 15.1 *Cases of Diabetes Mellitus Annually*

	52–207	12–51	2–11	<2	Nil	Total	
Cassava eaten	2	10	5	2	0	19	Full
Other staple							observed
foods	1	11	21	27	5	65	table
	3	21	26	29	5	84	

TABLE 15.2 *Cases of Diabetes Mellitus Annually*

	12–207	2–11	<2	Total	
Cassava eaten	12	5	2	19	Observed
Other staple foods	12	21	32	65	table
Cassava eaten	5·43	5·88	7·69	20	Expected
Other staple foods	18·57	20·12	26·31	64	table
	24	26	34	84	

$$X^2 = 16·9 \qquad n = 2 \qquad p < 0·001$$

reporting hospitals. (Hospitals which have failed to report on either variate are excluded.)

The chi-squared test suggests that a null hypothesis which postulates no relation between the distribution of cassava-eating and diabetes would only be justified once in a thousand. Whilst this is very strong statistical suggestion of a spatial association between the variates it is not, of course, proof of an aetiological connection. It can be claimed to be evidence sufficient to justify further research and to be a pointer to a specific line of inquiry. In such follow-up the possibility of a third and unknown further co-extensive factor should not be overlooked.

EXAMPLE B: CANCER OF CERVIX UTERI AND MALE CIRCUMCISION

Medical opinion on the hypothetical causative relationship between cancer of cervix uteri and male circumcision has been usefully summarised in two editorial articles in the *British Medical Journal*.[5, 6] From these circumcision would appear to be valuable chiefly as an aid to penile hygiene and the prevention of smegma collection.

Two lines of study are possible. Firstly it is possible to investigate the coital history of every woman patient with cervical cancer. It has been objected[7] that women may be genuinely unaware of the circumcision state of their various sexual partners. This line of inquiry may also be objectionable as interference in strictly personal matters.

Secondly, inquiries may be carried out amongst female populations whose coital experience may be regarded as homogeneous. Jewesses or nuns are examples of female populations assumed to be homogeneous in this respect.

Bantu Africa can provide similar examples. In certain tribes male circumcision is universal; in others it is never performed. It should thus be possible to obtain comparative evidence in areas such as central Africa where tribes of both customs reside.

Fig. 15.4 shows two centres where circumcision is practised. In Malawi circumcision is a religious rite amongst Muslims, who are estimated to make up variously from one eighth to three-quarters of the population. In North Western Zambia there is an area where circumcision (at age seven or eight) is universal. Outwards from this focus, fractions of the male population are reported to undergo circumcision. In part these fractions can be explained as being made up of migrant labourers attracted to mining towns from rural areas.

Fig. 15.3 shows the distribution of cases of uterine cervix cancer. Since the disease, if untreated, is invariably fatal, it is likely that most cases eventually present themselves in hospital even if suspicion and distrust of 'western' medicine cause delay in seeking treatment.

Cancer affects two sites in the uterus, the body (corpus) or neck (cervix). The survey specified both types together for report, but six hospitals mention that no cases of corpus cancer are seen and two others report it as extremely rare. For all practical purposes the map may be considered to show cancer of the cervix uteri.

Table 15.3 sets out the values of the variates, with the numbers in the cells again representing all the hospitals which have reported on both variates.

Since the purpose in this case is to study female populations whose

TABLE 15.3 *Cases of Cancer of the Cervix Annually*

	12–51	2–11	<2	Nil	Total	
All circumcised	1	4	2	4	11	Full
(a) 3 in 4	3	2	3	1	9	observed
(b) 1 in 2	1	1	1	–	3	table
(c) 1 in 4	2	7	–	–	9	
None circumcised	6	23	18	1	48	
	13	37	24	6	80	

TABLE 15.4 *Cases of Cancer of the Cervix Annually*

	2–51	<2	Nil	Total	
All circumcised	5	2	4	11	Observed
None circumcised	29	18	1	48	table
All circumcised	6	4	1	11	Expected
None circumcised	28	16	4	48	table
	34	20	5	59	

coital experience is homogeneous, the full table has been abbreviated in Table 15.4 to exclude hospitals which report that any fraction of local men are circumcised. That is to exclude lines 'a', 'b', and 'c'.

This leaves a reading for circumcision of either 'all' or 'none' and in the 'expected' values part of the table it can be seen by inspection that several individual cells fall below the minimum value of 5. The chi-squared test for significance cannot therefore be applied in this case. The evidence, for what it is worth, lies only in the discrepancy displayed between the expected and the observed values. This discrepancy is particularly marked in the 'Disease Nil' column of the table.

EXAMPLE C: CASSAVA EATING AND CIRCUMCISION

When the two geogen maps (Figs. 15.2 and 15.4) are compared by eye there is again a suggestion of similarity (see Table 15.5). Again it is necessary to combine columns before applying the chi-squared test. This has been done in Table 15.6.

TABLE 15.5 *Proportion of Males Circumcised*

	None	1 in 4	1 in 2	3 in 4	All	*Total*	
Cassava eaten	5	–	1	3	10	19	Full observed.
Other staple foods	44	11	2	8	–	65	table
	49	11	3	11	10	84	

TABLE 15.6 *Proportion of Males Circumcised*

	None	1 in 4–1 in 2	3 in 4–All	*Total*	
Cassava eaten	5	1	13	19	Observed
Other staple foods	44	13	8	65	table
Cassava eaten	11·1	3·2	4·7	19	Expected
Other staple foods	37·9	10·8	16·3	65	table
	49	14	21	84	

$$X^2 = 24\cdot75 \qquad n = 2 \qquad p < 0\cdot001$$

RESULTS

These three examples show the wide variety of result which may arise for interpretation and it is in this field that the greatest degree of inter-disciplinary co-operation is needed to ensure that suggestions are constructively followed up. A further point of value is that comparison of two variates is far less difficult to interpret than multi-variate regression, because each geogen is here considered in isolation.

Example C merely indicates that, to some degree, the tribal customs of circumcision and cassava-eating are co-extensive. It appears that, in the area considered, tribes whose men are circumcised happen to have cassava as their staple plant food.

Example B is slightly corroborative of the existing hypothesis which relates male circumcision and female uterine cancer. It makes clear that Africa may yield evidence and encourages the geographer to widen the field of data-collection so that statistically viable results emerge.

Example A seems to have the greatest intrinsic value. The term 'diabetes' covers a broad spectrum of medical conditions whose symptoms in common led to the diagnostic reports tabulated. The apparent relation here suggested with cassava may be the result (as in Example C) of both being common to certain areas or tribes. On the other hand, research in pathology or biochemistry might prove that the known toxic substances in cassava can act on the human body to produce either diabetes or a condition diagnostically simulating diabetes. Yet again the geographical correlation might arise from the effects of malnutrition disturbing the normal function of the pancreas.[8,9]

A computer can be used to compile tables from hospital data to show observed and expected values as above. Vast quantities of data can thus be rejected as showing correlational values too low to be of interest. On the positive side, however, each high value of correlation demands its own explanation. When that is not forthcoming it becomes a suggestion from medical geography for follow-up by other disciplines,[10] or for further and more detailed study using geographical methods of greater accuracy.

SUMMARY

An account is given of some aspects of a geographical pathology survey in parts of Central Africa. Examples are given of disease distribution maps for their intrinsic value in medical administration and for their possible correlation with maps of socio-environmental variables.

Whilst the many pitfalls are acknowledged, a simple method of comparing pairs of distribution maps is introduced and the examples are used to explain the interpretative problems which arise.

ACKNOWLEDGEMENTS

I wish most fully to acknowledge that this method of categories for surveying geographical variations of disease was introduced in a lecture to the South African Institute for Medical Research by Professor P. E. S. Palmer on 26th August 1963. My thanks are due to J. E. Kerrich, Professor of Statistics, and D. S. Henderson, Professor of Computer Science, both in the University of the Witwatersrand, for their kind assistance, and to the resident medical officers in central Africa for providing the data. Also to G. H. Adika, A.I.S.T., Chief Technician in the University of Zambia, who drew the maps. Errors of interpretation or understanding remain my own.

REFERENCES

1 ORDMAN, D. (1955) *S. Afr. med. J.*, **29,** 173.
2 WEBB, J. S. (1963) *The New Scientist*, **23,** 504.
3 MAY, J. M. (1954) 'Medical geography'. In P. E. James and C. F. Jones (eds) *American Geography: Inventory and Prospect*, p. 454. Syracuse.
4 GELFAND, M. (1966) *The Sick African*, 3rd. Edit. Juta Press.
5 *Br. med. J.* (1964) ii, 397.
6 *Br. med. J.* (1965) i, 1327.
7 OETTLÉ, A. G. (1961) *Acta Unionis Internationalis Contra Cancrum*, **17,** 915.
8 ZUIDEMA, P. J. (1959) *Trop. geogr. Med.*, **11,** 68.
9 SONNET, J., BRISBOIS, P. and BASTIN, J. P. (1966) *Trop. geogr. Med.*, **18,** 97.
10 DAVIDSON, J. C. *et al.* (1969) *Medical Proceedings (Johannesburg)*, **15,** 426.

16 Simple Chronic Bronchitis and Urban Ecological Structure

J. L. Girt

INTRODUCTION

Chronic bronchitis in Great Britain has been consistently associated with urbanisation and industrialisation. Howe[1] concluded from his maps of bronchitis mortality that regions of high mortality corresponded in the United Kingdom to the areas of dense industrial populations. His maps show how urban life is particularly conducive to the disease since the county boroughs throughout the country consistently stand out as areas of higher mortality amongst the more rural county areas immediately surrounding them. Since the disease is the third most important cause of death in males and the fourth most important in females, Morris suggests that 'chronic bronchitis, in particular, and respiratory disease more generally, are responsible for most of the historic urban/rural and North/South differentials (in mortality rates in Great Britain)'.[2]

It is not yet possible to describe a pattern of causation for the disease and its accompanying geographical gradient. Smoking has been consistently associated with the disease and there is less controversy over this postulated cause than over the disease's link with urban ways of life. Chronic air pollution has been suggested as a likely factor in this trend. The extent of the association between this factor and bronchitis is shown in Table 16.1,

TABLE 16.1 *Standardised Mortality Rates in Local Authority Areas in England and Wales Grouped by Inferred Level of Air Pollution*

	England and Wales	Rural Districts	Increasing Level of Air Pollution		
			County Boroughs		
			Clean	Intermediate	Dirty
Males	100	52	80	126	176
Females	100	59	71	132	186

where the standardised mortality ratios[3] in Britain in four types of environment based upon inferred levels of air pollution are given. Inhabitants of the most polluted and most industrialised urban areas are more than three times as likely to die from the disease as those living in rural areas. Other significant associations with urban/industrial environments have also been found. The disease is related to income and occupation, being particularly prevalent amongst the poorer sectors of populations.[4] These people tend to live in the most polluted parts of towns not only because of the proximity of the factory chimney to their residences, but, more important, the spatial concentration of domestic chimneys. The same, small working-class houses not only lead to high levels of chronic air pollution but also, through their poor construction, to damp living conditions and overcrowding. Thus the same people who are exposed to high levels of air pollution at home and at work are also exposed through overcrowding to upper respiratory infections to a greater extent than less crowded populations.[5] In addition, Turner[6] has postulated that overcrowding and dampness in houses may encourage the growth of certain fungi whose spores could act as a lung irritant. Respiratory infection is known to precipitate bronchitis, and lung irritation is also thought to be a likely aetiological factor. The coincidence of these factors, and many others which accompany them, is mainly responsible for the inability of science to isolate a specific pattern of causation for the disease. Is air pollution the cause of urban/rural, North/South mortality trends in Britain, or are they a result of other factors which spatially coincide with air pollution? It is extremely difficult outside controlled experimental conditions to ascertain this, since one cannot avoid inadvertently including the effects of excluded variables when examining the relationship between specific environmental factors and bronchitis. This chapter attempts to throw more light on this problem by studying the individual risks of having bronchitis and the resulting geographical distributions within one city, since, as we have hinted, it would appear that it is not the city but parts of the city which are probably responsible for these gradients in Britain.

PROBLEM AND APPROACH

Thus science has indicted city conditions for the high incidence of chronic bronchitis in Great Britain. This study was designed to estimate the basic spatial structure of simple chronic bronchitis, an early, incipient stage in the development of the disease, within a British city – Leeds, Yorkshire – and to attempt to derive a complex of factors which might be responsible for the generation of this pattern. Within a city the basic spatial, ecological structures are mainly based upon two trends – concentric growth and sectorial socio-economic differentiation. Simple chronic bronchitis appeared to be more closely linked with the second trend and showed marked sectorial morbidity trends. In fact within the city itself as great a range in prevalence was found as Morris found for mortality between urban and rural areas as reported in Table 16.1.

Chronic bronchitis, like most diseases, exists in many degrees of severity and it was not possible to investigate all the stages characteristic of the disease. Simple chronic bronchitis was chosen since it is thought to represent the initial stages in a process which, if left to continue, is usually irreversible. Thus, were it possible to prevent this stage, the more serious forms of the disease could be avoided. On the other hand, it must be remembered that the relationships found in this study do not necessarily hold for the other stages in the disease as well. Simple chronic bronchitis has been defined as 'the production of phlegm on most days for as much as three months in the year'.[7] The British Medical Council's Committee on the Aetiology of Chronic Bronchitis has developed a questionnaire which tests for this symptom in a series of simple questions.[8] These questions, with the addition of ones relating to smoking habits and residential and occupational histories, were put to a sample of the female population of Leeds. Those interviewed had been randomly selected from the total female population of Leeds, and, consequently, the bronchitics amongst them were found independently of whether they themselves were aware of this condition, and of whether they had sought treatment for it. In fact the onset of this condition is so insidious that many of those affected are unaware of its presence or its possible significance.

Disease is the result of interactions between the individual and his environment, and the risk of developing a disease will increase with lengthening exposure to a causal factor in the environment. Thus not only must a possible causal factor show spatial covariation with incidence of the disease but the risk of developing the disease must be linked to length of exposure. Both of these constraints were met by using individuals rather than areas as the basic unit of analysis. If spatial aggregates of people had been used it would not have been possible to measure the effects of length of exposure to variables on the spatial pattern of the disease. This also avoided the possible problems of interpreting ecological relationships which would have been found if spatial aggregates of people had been used as the main units of analysis.[9] A sample of females living in a representative range of environments in Leeds were interviewed to provide the data from which the spatial structure of morbidity in Leeds and the complex of environmental factors responsible for it was inferred. The latter was achieved by using the least squares regression technique to find the relationship between the risk of an individual having the disease and temporal exposure to the factor(s) in question. This relationship was then used, through a simulation method, to estimate the spatial pattern of morbidity such factor(s) could be expected to produce. In the conclusion a preliminary attempt is made to relate the significant relationships found to spatial suburbanisation and socio-economic processes within the city.

THE DATA

Leeds, with a population of approximately one-half million, is the regional capital of West Yorkshire. It is a town with very apparent nineteenth-century industrial foundations based upon the clothing, engineering and chemical industries. The town is divided latitudinally by the River Aire whose banks are almost totally occupied by industry, which also extends from the centre of the town southwards towards a coalfield immediately to the south of the town. The northern half of the city, apart from two large clothing factories, is primarily residential, the southern half is a mixture of industry and housing. A high proportion of the houses were built in the last century and many back-to-backs* remain in the central and industrial areas. The ecology of the city appears to be such that Mann's model of the spatial, ecological structure of a British city[10] would appear to be applicable with little modification. The only modification required is rotation. By rotating the structure so that the middle- and upper-class sector is in the north of the city rather than in the west the models become a reasonable representation of the situation in Leeds. It is in this revised form that the model is shown in Fig. 16.1. The modification is probably the result of the attractiveness to industry not only of the east–west flowing Aire and its valley but also the coalfield to the south.

It was decided to interview between 700 and 800 residents contained within the city limits of Leeds. In order that a representative range of environments would be studied, thirty areas or quadrats were systematically selected and the occupants at twenty randomly selected residences with each approached for interviewing. It was felt that either the west or east half of the city would contain the required range of environments and, for interviewing convenience, the eastern half was chosen. A study area was defined by the city boundary on the north, east and south, and in the west by the northing line on the Ordnance Survey 1:25,000 map of the area that came closest to the city centre. This method of division was used since the Ordnance Survey map also formed the basis of a convenient method of defining sample areas or quadrats within the study area. A sample of people had to be chosen which would be representative of the populations living in a representative range of environments in Leeds. At the same time a

---

* Back-to-back housing consists of terraces constructed two rooms deep and two or three storeys high. Each house is one room deep and two or three storeys high. There are, thus, no back yards and often toilet facilities are entered from outside the house and may be shared by a number of houses. In Leeds these houses have been observed at densities of more than 6,800 to the square kilometre.

means had to be found of ascertaining the addresses of those who were to form the sample population. The coordinate grid on the 1:25,000 map defined a series of cells over Leeds each one half kilometre square and covering the territory of four Ordnance Survey 1:2,500 plans. The plans had been recently revised in 1966 when the study began and provided the address of every residence shown in the area. Thus the 1:25,000 map was

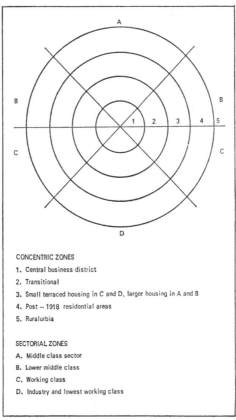

CONCENTRIC ZONES

1. Central business district
2. Transitional
3. Small terraced housing in C and D, larger housing in A and B
4. Post – 1918 residential areas
5. Ruralurbia

SECTORIAL ZONES

A. Middle class sector
B. Lower middle class
C. Working class
D. Industry and lowest working class

FIGURE 16.1 The ecological structure of a British city (Leeds).

used to define 1:2,500 plans or combinations of plans which formed the sample quadrats. A variation in size of quadrat was adopted since the areas covered by different types of housing increase as one moves away from the centres of cities, and otherwise quadrats near the periphery would have small populations and many quadrats there would be essentially similar both in physical and socio-economic structure. The territory covered by an individual 1:2,500 plan was combined with contiguous territories if its population was insufficient for the land use of the area to appear from the

1:25,000 map to be more than 40 per cent residential (excluding gardens). Completely non-residential territories were omitted from the study area. Territories more than 40 per cent residential were made into individual quadrats. In all, ninety quadrats were defined. Thirty of these were systematically selected to form the sample environments in the study. They ranged in size from individual territories of one-quarter kilometre squares to contiguous areas of three square kilometres (see Fig. 16.2). Within each sample quadrat a sample of residences were randomly chosen from which the sample population was to be drawn.

It was decided to limit the study to females. Resources being limited, and it being an exploratory study into a relatively new field, it was thought that females would be easier to study since their occupational histories would be simpler and shorter than males. This would help to reduce the complexity of relationships involved in the analysis of the data. In addition, a better interviewing response rate could be expected from females than from males during the day. Oswald[11] and the College of General Practitioners[12] have both concluded that the apparently higher incidence of bronchitis in males than females was probably due to different exposures to aetiological factors, and that sex was not a direct aetiological factor in itself. However, no information is available to show how relevant the conclusions of this study could be to males, though generally it has been found that the prevalence from the disease is lower amongst females than males at the same time as both sexes exhibit similar geographical and socio-economic gradients.[13]

It was also decided to limit the study to those aged fifteen and over since bronchitis is primarily a disease of middle and old age.

In each sample quadrat the female occupants of twenty randomly selected residences were interviewed resulting in a total sample population of 733 with a response rate of 93 per cent. Each respondent was questioned to ascertain whether she had the symptoms of simple chronic bronchitis and asked additional questions regarding her smoking habits and occupational and residential histories. (See below.)

THE SIGNIFICANCE OF THE SPATIAL PATTERN

Fig. 16.2 shows, in addition to the location in Leeds of the sample quadrats, the spatial variations in sample prevalence of simple chronic bronchitis found in the interview survey. It is the purpose of this section to show how it was ascertained that the pattern contained significant variation. This was achieved by making use of the Poisson distribution. Under a Poisson process any individual would have the same probability of having bronchitis and any differences in prevalence between the quadrats would be solely due to chance.

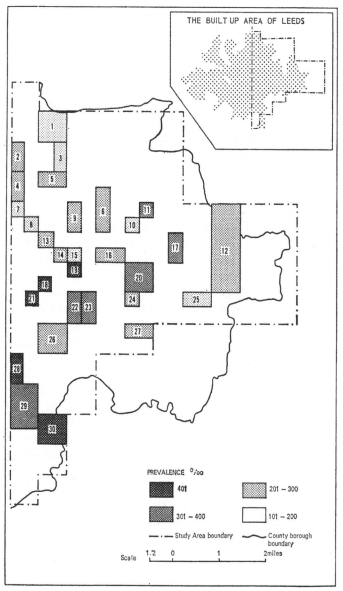

THE BUILT UP AREA OF LEEDS

PREVALENCE º/oo

401

301 — 400

201 — 300

101 — 200

— · — Study Area boundary

County borough boundary

Scale   1.2   0   1   2miles

FIGURE 16.2 Sample prevalence of simple chronic bronchitis.

Given p the mean probability of occurrence of bronchitis,* the probability of there being exactly k bronchitics in a quadrat is given by:

$$p(k;\Lambda) = \frac{e^{-\Lambda}\Lambda^k}{k!}$$

where $\Lambda = np$ or the number of bronchitics expected in that quadrat with a constant probability of occurrence, and $n =$ the number of individuals in the sample population of that quadrat.[14]

There is a less than 0·05 probability of the extreme prevalences in the quadrats occurring through a Poisson process and, hence, it was concluded that significant spatial variation in bronchitis prevalence did exist between the places included in the sample.

LINKS WITH MODELS OF URBAN ECOLOGY

Two apparently basic dimensions of urban ecology are built around the concepts of Burgess and Hoyt, namely that a city expands and is organised around concentric and sectorial zonation and growth, and it is the purpose of this section to link the spatial pattern of bronchitis morbidity in Leeds to these two dimensions. Cities expand around a centre in a concentric fashion. Thus, for example, age of housing tends to be distributed concentrically. Inhabitants of a city tend to be distributed into sectors according to socio-economic status around the centre, each sector growing concentrically outwards as the city grows. Socio-economic status then tends to vary according to the angular measurement about the centre of the city. Murdie[15] has shown how indices derived from census data in Toronto are distributed according to this scheme. Type and density of housing in Leeds, as one would expect from Mann's model, itself derived from the work of Burgess and Hoyt, tends to vary concentrically around the city centre. Since air pollution is related to density of housing it too will show a similar tendency. If air pollution does cause bronchitis the disease should show a similar spatial pattern. On the other hand, if bronchitis is more a result of socio-economic factors than the contemporary physical environment of the city one should expect a sectorial tendency in the distribution of morbidity. In addition, a sectorial pattern of bronchitis morbidity could be produced by suburbanisation processes. Inhabitants of the sectors of low socio-economic status who now live in the suburbs of cities are often occupying local authority housing having been moved from more central areas under slum clearance schemes. Consequently, they may have been exposed to similar or even poorer conditions of housing and pollution for a considerable part

---

* The mean probability of having simple chronic bronchitis in the sample population was 0·294. This figure will not be characteristic of the whole female population of Leeds as certain types of environment probably have been over-represented in the sample.

of their lives as those still living in central areas and may, as a result, have as high an incidence of bronchitis.

Mann's model of the ecology of a British city would thus appear to be a useful device to describe the pattern of morbidity whilst at the same time relating the pattern to broad spatial trends in the city. Thus it was found that the prevalence of bronchitis in the quadrats was significantly related to both the distance in miles from the city centre to the centre of each quadrat, and also the angular measurement of the quadrat centre to the centre of the city. The following regression equation gave the best fit:

$$X = 0.253 - 0.034 \text{ Distance} + 0.002 \text{ Angle}; \quad R = 0.73 > 0.001$$

For diagrammatic purposes it is more convenient to divide the city into sectors and to examine the prevalence within each sector rather than use angular measurements. Thus, four equal-sized sectors were defined covering the study area and radiating from the city centre. It was found, using the Poisson test, that the prevalence of bronchitis in the two middle sectors was not significantly different and, consequently, these two sectors were combined. Then the following regression equation was found:

$$X_1 = 0.501 - 0.034 \text{ Distance} - 0.220 \ X_3 - 0.109 \ X_4; \quad R = 0.073 > 0.01$$

where $X_3 = 1$ if the quadrat was in the northern sector (A), 0 if not, and $X_4 = 1$ if the quadrat was in the central sector (B), 0 if not. The predicted pattern according to this equation is shown in Fig. 16.3 where, in addition to the sector boundaries, concentric zones are drawn at one half-mile intervals. The equation has been used to infer the pattern for the whole city. The predicted prevalences refer to the maximum that could be expected in each zone according to the regression equation.

Fifty-three per cent of the variance in bronchitis prevalence between the quadrats is accounted for by either of the two regression equations. Although the contribution of the distance variable in each case is small in relation to the sector variables, all contribute significantly to the overall variation. The equations suggest that the spatial pattern of morbidity in large British cities will show both concentric and sectorial trends around the city centre. Bronchitis morbidity will decrease as one moves from working class industrial sectors towards middle class areas. Thus in Leeds the southern sector (C) has the highest predicted proportion of bronchitics and contains the majority of quadrats in the study with heavy industrial environments and with predominantly lower class residents. Sector A in the north of the city with the lowest predicted proportion of bronchitics contains a high proportion of privately owned, large Victorian and twentieth-century houses with a high proportion of middle class residents. Sector B, the central sector, falls between the other two sectors in predicted morbidity and contains all forms of housing from the dense back-to-backs and semi-detached council housing estates of sector C, to the areas of post-1918 and semi-detached private housing of sector A. It is surprising that the concentric component did not prove more significant; that it is both relatively and absolutely small compared to the sectorial effect suggests that little effect can be attributed to present housing environments. The whole of the central area in all three sectors is congested and polluted consisting pre-

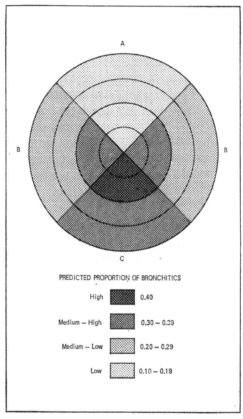

FIGURE 16.3 Predicted prevalence of simple chronic bronchitis from the ecological model.

dominantly of pre-1900 housing, much of it back-to-back, or terraced. It contains, particularly in the south, much of the heavy industry of Leeds. Here one would expect air pollution levels to be their highest, yet there is only a small gradient in prevalence towards the suburbs. In the suburbs, in all three sectors, semi-detached and detached post-1918 housing is dominant. Thus it would appear that, within the city, socio-economic differentiation is of more significance to current levels of simple chronic bronchitis than present-day conditions relating to age and quality of housing and environment. It must be remembered that this may in part be the result of large-scale suburban growth in the last twenty or so years which has led to an outward migration of people from the centre of the city, particularly from sectors B and C, to local authority housing estates, particularly in sector C.

Although only based upon the experiences in a single city, the widespread generality of the concentric and sectorial trends in urban ecology leads one to suggest that Fig. 16.3 is a viable descriptive model of the ecology of bronchitis in a large, industrial British city. It now remains to suggest a process whereby such a pattern could be created. This problem proved exceedingly difficult to solve and the rest of this paper must be regarded as only an initial conceptualisation of the factors and associations involved.

CRITERIA FOR EVALUATING THE ROLE OF FACTORS IN
THE SPATIAL DISTRIBUTION OF A NON-CONTAGIOUS DISEASE

Before examining the possible effects of particular factors on the spatial distribution of bronchitis in Leeds it is necessary briefly to review the way in which such effects can be detected and measured. Three conditions must be fulfilled before a factor can be considered to cause a disease. Consistent relationships must be found between exposure to the factor and the disease in real life; similar relationships must be shown to occur under controlled laboratory conditions; and the removal of the factor from the environment must prove efficacious.[16] With a disease, such as bronchitis, of which very little is known beyond the first condition, it will obviously be very difficult to justify conceptualising significant statistical relationships in dogmatically causal terms. Rather, relationships found, especially when consistent with conclusions from other studies, should be regarded as hypotheses which require confirmation or rejection from future research,[17] particularly that in the laboratory. Such a stage is beyond the realm of the geographer. He is, however, likely to be able to play a useful role in planning the elimination of a factor from the environment. It is thus important to relate significant factors wherever possible to broader processes involved in human existence. This is why the links with models of urban ecology are so interesting.

Since there is a great deal of interconnection between environmental factors, it is very difficult for the geographer to be able to achieve more than the definition of groups of factors which may be or which definitely are not related to bronchitis in a given range of environments. Many of the single factors in themselves are often complex, containing perhaps only a single component which directly affects morbidity.

Whether a variable represents a specific factor or a complex of factors does not affect the fact that, if the relationship being examined does have causal links with chronic bronchitis, consistent positive relationships will be found between length of exposure and the intensity of exposure, and the individual's risk of having the disease. In the study in cases where the factor involved could vary in intensity, for example, air pollution, a number of variables were defined each representing a particular level of intensity. The relationships between length of exposure to variables and the individual risk of having bronchitis were represented by least squares regression equations. A dummy variable was used to indicate the presence or absence of simple chronic bronchitis in each individual in the sample. This variable formed the values for the dependent variable in equations which gave the predicted conditional probability of having bronchitis given exposure to the other variables in question.

The actual numbers of bronchitics that could be expected in each quadrat as a result of a particular regression model, excluding the intercept function, were derived by a simple Monte Carlo simulation technique. The predicted conditional probability excluding the intercept function of having the disease was found for every member of the sample, and these values were compared with random numbers. The intercept function was not included since it did not measure the effect of exposure to the variables in the model on the risk of having bronchitis. Those with conditional probabilities equal to or less than the random numbers were taken to be bronchitic as a result of the model. The proportion of simulated bronchitics were compared with the actual proportion in each quadrat in order to ascertain whether the regression model in question was likely to contribute to the real world spatial distribution of the disease.[18]

The effects of air pollution, overcrowding, damp housing, type of occupation and smoking were analysed this way. The findings will be discussed in two groups, first, that relating to air pollution, and second, that relating to all the other factors.

AIR POLLUTION

Much has been written about the apparent effects of air pollution on bronchitis. Generally, the acceptance of the link between the two has been based upon the fact that the prevalence of the disease is highest where air pollution levels are also highest. However, this relationship did not appear to apply within Leeds to the same extent that it applies at larger geographic scales, for example, between local authority areas in Great Britain. Within Leeds, although there was a concentric trend in the bronchitis prevalence rates for the quadrats, the predominant trends was sectorial. One would expect air pollution levels to show a concentric trend because they are related to density of housing and industry. No adequate direct source of air pollution levels in the study area was available, but using inferred levels of pollution, no relationship could be found between length of residence in a quadrat and the risk of having bronchitis.

From air pollution data from Sheffield as well as Leeds, it was found that both smoke and sulphur dioxide levels were related to density of housing.[19] In Leeds, using more limited data than for Sheffield, a significant relationship was also found between sulphur dioxide pollution and elevation. Both of these relationships were expected. The density of domestic chimneys, the main source of smoke pollution, naturally bears a close relationship to housing density. Sulphur dioxide pollution, on the other hand, is produced in large quantity by industry, particularly heavy industrial plants.[20] These, because of their age, and also the need for waterside locations, tend to be found at the lower altitudes within cities.

However, whether the sample population was analysed as a whole or in groups by density of housing in their local area or elevation of their home quadrat, no relationship could be found between length of residence in the quadrat and the risk of having bronchitis. This finding should not be construed as indicating that air pollution is not a factor affecting the prevalence of bronchitis within Leeds or larger geographical units. Only one stage in the disease has been studied this way and air pollution could well be a factor influencing degeneration to more severe stages. No mention has been made of the possible effects of air pollution at work or in previous environments. Some conclusions regarding these factors will be made in the next section.

THE REMAINING FACTORS

All the other factors reported in this study were significantly related to bronchitis. The factors referred to smoking habits, past and present living conditions, environmental conditions in past residential areas and some aspects of occupation.

Smoking habits were measured by the number of years each individual had smoked.

$$Y = 0.222 + 0.010 \text{ Smoking}; r = 0.25$$

Whilst not a dominant factor, it did produce a significant sectorial trend in simulated prevalences ($F = 18.06 : p > 0.01$, where p = level of significance) with a gradient of increasing prevalence from sector A to sector C. There is a large body of evidence linking smoking with bronchitis,[21] but what is surprising is the extent to which the habit shows a spatial distribution, since it has usually been assumed that the habit is fairly evenly spread throughout all social classes.

Present living conditions were represented by three variables. Two measured density of occupance of living rooms in the present residence;

the first, referred to as O_1, comprised the number of years spent in the present residence at a density of more than one but less than two persons to a living room. The second, referred to as O_2, comprised the same for densities of two or more persons per living room. The spread of upper respiratory infection, a precursor of bronchitis as well as the disease itself, has been found to be related to such variables.[5] A measure of the years spent living in damp housing formed the third variable. Overcrowding, by raising air humidity, and dampness would encourage the growth of fungi of the species *aspergillum* and *cladosporium* which Turner[6] has suggested could produce respiratory catarrh as a result of an allergy to their spores. The relationship was as follows:

$$X = 0.268 + 0.000\ O_1 + 0.000\ O_2 + 0.011\ \text{Dampness};\ R = 0.16$$

The independent variables do significantly contribute to the overall variation in the dependent variable ($F = 6.68 : p > 0.01$). It would appear that Turner's hypothesis regarding the effect of fungi may be realistic, but that density of occupation of present residences is not important. The relationship did not simulate a pattern which resembled the actual pattern at all closely, nor were the simulated prevalences related in any way to distance from the city centre and angle from the centre.

Past living conditions, on the other hand, proved to be very significantly related to the risk of having bronchitis. The same variables were used as for present living conditions but now they referred to all residences prior to the contemporary one.

$$X = 0.156 + 0.004\ O_1 + 0.005\ O_2 + 0.006\ \text{Dampness};\ R = 0.25$$

Not only is dampness, but also the two density of occupation variables are of significance. The simulated prevalences the model produces are compared to the actual prevalences in each quadrat in Fig. 16.4. The dotted lines on the graph indicate the limits within which actual and simulated rates only differ through chance ($p = 0.95$).[22] Eight quadrats fall outside these limits and obviously the goodness of fit leaves something to be desired. Again, simulated prevalences did not show any significant sectorial or concentric trends. However, from the degree of fit shown in Fig. 16.4, it would appear that these variables do contribute to the overall spatial pattern of prevalence. The regression coefficients are consistent with both Turner's hypothesis and the one relating overcrowding to respiratory infection and bronchitis.

In order to ascertain the possible significance of air pollution in previous areas of residence, information was gathered on the types of environment respondents had lived in prior to their present one. They were asked to allocate each residence they had lived in to one of four categories. 'Enclosed dirty' was defined as polluted, densely housed parts of towns of similar character to Hunslet (quadrat number 28 in the study) and similar areas in Leeds. Usually such areas consisted predominantly of back-to-back houses. 'Enclosed clean' referred to all those terraced urban environments which could not be put in the former category. 'Open' referred to all other urban environments and the fourth category, 'rural' is self-explanatory. It was hoped that these categories used as variables would measure relative levels of air pollution in previous residences. The classification scheme was tested to see if residents would consistently classify similar environments in the same category. Each respondent was asked to allocate an environmental type to her present residential area. There was a large

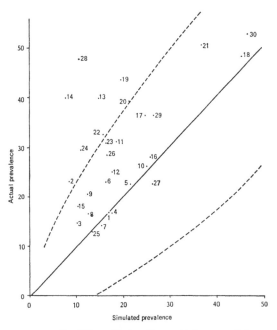

FIGURE 16.4 The effects of past living conditions.

measure of agreement as to the appropriate type by the sample population of each quadrat, and the quadrats were assigned a type consistent with the criteria used to define each category.

This group of variables is closely related to living conditions and, consequently, a close relationship was found between length of exposure to these variables and the risk of the individual having bronchitis.

$$X_1 = 0 \cdot 142 + 0 \cdot 008 \text{ Encl. dirty} - 0 \cdot 002 \text{ Encl. clean} - 0 \cdot 007 \text{ Rural}; R = 0 \cdot 27$$

This equation also reproduced a simulated pattern of prevalence which was poor approximation to the real life one (see Fig. 16.5). Discussion of the role of these variables will begin a little later. Let us now turn our attention to the final set of variables examined, those relating to occupation.

Air pollution at work as well as in the residential area could be a causal factor in bronchitis. Although many people lived close to their workplace, it was decided to construct variables to measure this separately. One variable, referred to as 'industrial', measured the number of years each individual had worked in a heavy industrial plant or in an industrial area. A second variable measured the same for work in city centres where pollution levels are also likely to be high. The final variable in this group, abbreviated to 'workfloor', measured the time each respondent had worked on machinery fabricating a product which could involve dust. In Leeds, this applied to a high proportion of women, the clothing, paper and chemical industries

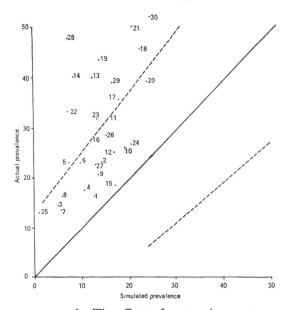

FIGURE 16.5 The effects of past environments.

being large employers of female labour. Lung irritation through exposure to dust is a widely recognised factor in the possible aetiology of chronic bronchitis. Once again, a significant relationship was found with the risk of having bronchitis.

$$X_1 = 0.226 + 0.002 \text{ Industrial} + p.002 \text{ Commercial} +$$
$$0.007 \text{ Workfloor}; R = 0.21$$

A strong sectorial trend was found in the simulated prevalences from this equation ($Y = 0.107 + 0.002$ Angle; $r = 0.52 : p > 0.01$).

With the exception of smoking, all the variables included in this section have been highly interrelated and one cannot be specific as to the actual factors each group of variables represents. Because of this, it was decided to combine the groups together in a step-wise type of regression programme. This began with the smoking model since smoking was not significantly related to other independent variables. The other groups of variables were successively introduced into the analysis but only allowed to enter the model if their contribution to the overall sum of squares was significant at the 0.05 level on an F test.[23] Present living conditions were the first group to be examined and it was found that they did make a significant contribution ($F = 16.76 : p > 0.01$). Past living conditions were the next group to be introduced and they too proved to be significant ($F = 13.00 : p > 0.01$). If, unlike under present conditions, air pollution in the past was a factor, it would be reasonable to expect that the contribution of past environmental conditions should prove significant after the effect of past living conditions had been included in the analysis. The contribution was significant ($F = 5.85 : p > 0.01$) and the resulting equation is given below:

$$X_1 = -0.002 + 0.009 \text{ Smoking} + 0.007 \text{ Encl. dirty} - 0.002 \text{ Encl. clean} - 0.003 \text{ Open} + 0.006 \text{ Rural} + 0.001 \text{ Past } O_1 + 0.001 \text{ Past } O_2 + 0.004 \text{ Past Damp} + 0.002 \text{ Present } O_1 + 0.003 \text{ Present } O_2 + 0.010 \text{ Present Damp};$$
$$R = 0.42$$

Air pollution perhaps could be the factor responsible for the significant regression coefficient for 'enclosed dirty' were it not for the coefficients for 'enclosed clean', 'open', and 'rural' as well. Air pollution cannot be correlated with 'enclosed dirty' and 'rural' environments at the same time. In fact, one wonders what role rural–urban migration has played in urban diseases from the size of its effect in the regression equation.

The occupational variables were then introduced but proved to have no significant effect ($F = 1.19 : p < 0.05$) and were dropped from further consideration.

Age has been deliberately neglected in this paper as a factor in the aetiology of bronchitis since it is felt that it is not a causal factor. It does have a significant relationship to the risk of having bronchitis ($Y = 0.003 + 0.006$ Age; $r = 0.28$), but this appears to be more a result of age being representative of length of exposure to causal variables than a direct agent in itself. This conclusion is confirmed by the fact that when age was introduced as a factor to the equation given above in the stepwise regression programme, its contribution to the overall sum of squares was insignificant ($F = 2.96 : p < 0.05$).

We are left with a final model which contains variables relating to past and present living conditions, past environmental conditions and smoking. It would not appear from analysing this equation and from previous conclusions that air pollution is a factor of significance in the aetiology of simple chronic bronchitis. Not only was no relationship found between pollution in present residential areas and bronchitis, but the effects of the past environmental variables in the final equation are not as one would expect if the factors were related. In addition, no occupational effect could be discerned above that possibly already represented in the other variables. Chronic bronchitis from the results of this study, at least, would appear to be more related to what goes on inside the home than outside it.

The final model produced a simulated pattern of prevalence which closely resembled the real-life pattern (see Fig. 16.6), only one area having

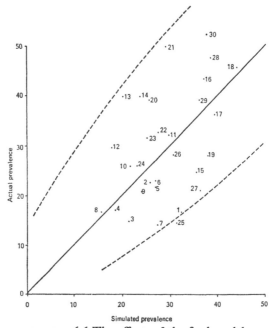

FIGURE 16.6 The effects of the final model.

a simulated prevalence significantly different from an actual one. A significant sectorial but no significant concentric trend was found in the simulated prevalences ($Y = 0.242 + 0.004$ Angle; $r = 0.36$).

CONCLUSIONS

We have shown how bronchitis prevalence in Leeds can be related to an ecological model of a British city. The disease shows a distinct sectorial gradient in prevalence and a less pronounced, but still significant, concentric one. Present levels of air pollution, and possibly past levels as well, are not related to the risk of having the disease. The final regression model combining effects of smoking, past and present living conditions, and past environments, gave a good estimation of the actual pattern of prevalence although it did not simulate a concentric tendency in the prevalence of the disease. Its high degree of fit, however, suggests that it is a useful construct and the lack of a concentric trend may be the result of the random element in the simulation model rather than something inherent in the regression equation. From the model we can lay the blame for the high incidence of bronchitis in Leeds amongst females on housing conditions and smoking. The disease probably shows a strong sectorial component since it is related to a degree of overcrowding and dampness in housing. Both of these will be particularly common today and in the past amongst the lower income groups. The prevalence of the disease is markedly higher in the lower income sector than elsewhere. It would appear that the disease is largely the result, excluding the effects of smoking, of small, inadequately constructed houses which, particularly in Victorian times, were produced in quantities to house the working man and his family. Rural conditions also appear to have played a similar role. Today these effects are noticeable even in the suburbs. This may be due to contemporary conditions there or due to sectorial outward movements of people or to both. The relatively large regression coefficients for the two variables measuring length and time spent in damp housing, together with the significant effects of 'enclosed dirty', 'rural' and the variables measuring density of occupation, suggest that Turner's hypothesis is a realistic one and that overcrowding, particularly when accompanied by dampness, is very conducive to the disease. Such conditions would be particularly common in the countryside and in the less attractive, over-occupied and over-industrialised urban areas represented by the 'enclosed dirty' variable.

We must hope that suburbanisation, rising living and housing standards and the tendency to have smaller families will be continued for bronchitis to be beaten. Obviously it is not just a question of slum clearance, although this is a positive step, for it reduces the amount of 'enclosed dirty' environments in existence, but a question of minimising overcrowding and maintaining a high standard of construction as well. In Leeds, at least, and probably in most other cities as well, the solution lies in the hands of the local authorities, since it is they who are responsible for most of the home construction now taking place in low income sectors such as sector C. With all these remedies the disease will still remain a problem, however, until smoking ceases to be a popular habit and those that partook of the habit are gone, but it is difficult to see how this problem can be related to the city.

REFERENCES

1 HOWE, G. M. (1963) *Atlas of Disease Mortality in the United Kingdom*, pp. 54 ff. London.
2 MORRIS, J. N. (1967) *Uses of Epidemiology*, p. 205. Edinburgh.
3 Standard mortality rates take into account variations in age-sex structures between groups of people which can lead to differential absolute mortality rates. For the method of calculation see Howe, G. M. (1963) *op. cit.*, pp. 3–6.

4 See, for example, OGILVIE, A. G. and NEWELL, D. J. (1957) *Chronic Bronchitis in Newcastle-upon-Tyne*, Edinburgh; HIGGINS, I. T. T. *et al.* (1959) 'Population studies of chronic respiratory disease: a comparison of miners, foundry-workers and others in Staveley, Derbyshire', *Brit. J. industr. Med.*, **16**, 255.

5 CRUICKSHANK, R. (1958) 'A survey of respiratory illness in a sample of females in London'. In: J. Pemberton and H. Willard (eds) *Recent Trends in Epidemiology*. Oxford.

6 TURNER, W. C. (1964) 'Air pollution and respiratory disease'. *Proc. Roy. Soc. Med.*, **57**, 618.

7 FLETCHER, C. M. (1959) 'Chronic bronchitis: its prevalence, nature and pathogenesis'. *Am. Rev. resp. Dis.*, **80**, 264.

8 Medical Research Council (1962) *Instructions for the Use of the Short Questionnaire on Respiratory Symptoms* (1960). London.

9 This problem of interpretation of ecological relationships has received little attention from medical geographers. Some of the problems involved in making causal inferences from aggregated data are discussed in BLALOCK, H. M. (1964) *Causal Inferences in Non-experimental Research*, Chapel Hill, pp. 94–8. See also the first part of DOGAN, M. and ROKKAN, S. (eds) (1969) *Quantitive Ecological Analysis in the Social Sciences*. Cambridge, Mass.

10 MANN, P. H. (1965) *An Approach to Urban Sociology*, pp. 95 ff. London.

11 OSWALD, N. C. (1958) *Recent Trends in Chronic Bronchitis*. London.

12 College of General Practitioners (1961) 'Chronic bronchitis in Great Britain'. *Brit. med. J.*, ii, 973.

13 See, for example, HIGGINS, I. T. T. and COCHRANE, A. L. (1961) 'Chronic respiratory disease in a random sample of men and women in Rhondda Fach in 1958'. *Brit. J. industr. Med.*, **18**, 255.

14 HALD, A. (1952) *Statistical Theory with Engineering Applications*, p. 714. New York.

15 MURDIE, R. 'Factorial ecology of metropolitan Toronto, 1951–61'. University of Chicago, Dept. of Geography, Research Papers, No. 116.

16 SPAIN, D. M. (1960) 'Discussion: sociocultural factors in chronic organic disease'. *Ann. New York Acad. Sci.*, **84**, 1031; PRINDLE, F. A. (1968) 'Environmental health: clinical and epidemiological considerations'. *Archs environ. Hlth*, **16**, 69.

17 MCGLASHAN, N. D. (1969) 'The nature of medical geography'. *Pacific Viewpoint*, **10**, 60–4.

18 For an extensive use of this type of Monte Carlo simulation technique see ORCUTT, G. H. *et al.* (1961) *Microanalysis of Socioeconomic Systems*. New York. In order to minimise the variance of the simulated prevalences, the sample size was artificially increased fourfold by repeating the simulation process four times for every individual.

19 The Sheffield data is given in PEMBERTON, J. *et al.* (1959) 'The spatial distribution of air pollution in Sheffield'. *J. Air Pollution*, **2**, 175. The Leeds data was made available by Leeds Department of Health.

20 Minister of Technology (1963) *Warren Springs Laboratory Notes on Air Pollution, No. 3, Airborne Pollution*. Stevenage.

21 See, for example, HIGGINS, I. T. T. (1957) 'Respiratory symptoms, bronchitis and ventilatory capacity in a random sample of an agricultural population'. *Brit. med. J.*, ii, 1198; HOLLAND, W. W. *et al.* (1965) 'Respiratory diseases in England and the United States', *Archs environ. Hlth*, **10,** 338.

22 The formula for these limits can be found in YAMANE, T. (1964) *Statistics, an Introductory Analysis*, pp. 497 ff. New York.

23 JOHNSTON, J. (1963) *Econometric Methods*, pp. 123-6. New York.

17 Leukaemia and Housing: An Intra-Urban Analysis

G. E. A. Dever

INTRODUCTION

On rare occasions leukaemia occurs in two or more occupants of the same house. The significance of such 'leukaemia houses' is hard to assess. While they may represent chance events, it is possible that residence in certain houses somehow predisposes the occupants to leukaemia.

(HEATH, 1969)

LEUKAEMIA AND HOUSING

Diseased leukaemia patients obviously lived in housing units of some kind. It is possible, therefore, that the housing unit may be a factor which enhanced the transmission of the disease. If multiple cases of leukaemia were found in unrelated individuals living in the same housing unit, then the hypothesis of a relationship between leukaemia and housing would be strengthened. Likewise, if leukaemia occurred in victims who had lived next door to each other, the same hypothesis would be substantiated. Given the stated aetiology hypothesis it is interesting to note that Heath has drawn attention to a case where four persons associated with the same house had contracted leukaemia over the eighteen-year period between 1950 and 1967.[1] The house is a single dwelling constructed of concrete blocks and wood and located in a residential section near the centre of town. It has three bedrooms, a living-room, kitchen and a bath. It has city water and is heated by space heaters. As Heath points out, 'dwellings such as the one described may well be logical places to look for possible associations between the house and leukaemia'.

In addition to Heath's study, three cases of Hodgkin's disease and three cases of leukaemia have been reported in two adjoining houses (A and B) in a town in south central Pennsylvania.[2] House A is brick and house B is frame. Both houses have two storeys, an attic and a basement. House B had hot air from a coal furnace but switched to gas hot air in 1954. House A was divided into two units which had heat by coal hot air in one unit (this unit developed the leukaemia case) while the other was heated by gas hot air.

Both had excellent sanitary facilities. Gilmore concluded that 'environmental aspects appear to be the most logical explanation'.

These two studies point up a major area for research which can be specified in the question: 'Are the housing units in which leukaemia occur any different from those homes of a non-diseased population?' Since it cannot be determined from two studies or three homes if there are common factors to each house, a thorough study of leukaemia patients and their homes will now be investigated. The study will also attempt to determine the proper spatial scale for the identification of risk areas.

THE ANALYSES OF HOUSING WITH VARIOUS SCALES

In the remainder of this paper, four different scales and several housing variables will be analysed. It can be expected *a priori* that a relationship between the housing variables and leukaemia will vary depending on the scale of investigation. The four scales to be investigated are: (1) the census tract; (2) the block; (3) the tax district; and (4) the individual housing unit.

THE CENSUS TRACT SCALE

The census tract analysis represents the aggregation of data into seventy-five different tracts. Table 17.1 gives the results of the simple and multiple regression and correlation analyses.

The table indicates that all the nine variables are extremely weakly correlated with the leukaemia incidence rates. Variable 9, average rent, has a correlation coefficient of ·3792 in the simple equation, while it is ·3318 in the multiple case. Both of these values are significant at the ·01 level. This correlation suggests two interpretations, namely either that the rent paid by an individual is a substitute measure for his socio-economic status or that the apartment is in good condition. In order to choose between these two interpretations, it is necessary to discuss the influence of other variables. Thus, it should be noted that the simple equation shows both average value of the home and average number of rooms in the owner-occupied units to be positively and significantly related to the dependent variable.

The analysis indicates that an increase in the number of rooms per unit is associated with an increased incidence of leukaemia. This can be interpreted as contradicting the hypothesis that a viral disease is best transmitted in a crowded home or under crowded conditions. One may note that variable 12, which indicates the degree of crowding in the house, is not significantly correlated with the incidence of leukaemia, although it has a negative regression coefficient. Furthermore, the positive sign of the partial correlation coefficient for variable 10, the average number of rooms in the renter-occupied units, suggests that increased leukaemia rates are associated with more rooms per unit.

TABLE 17.1 *Census Tract Scale Correlation and Regression Analysis*

Variable	Var. no.	Simple Regression		Multiple Regression	
		Regression coefficient (b)	Correlation coefficient (r)	Regression coefficient (b)	Maximum partial correlation coefficient (r)
Per cent of housing units sound	2	·0696	·2177	·0135	·0203
Per cent of housing units with plumbing	3	·1346	·1346	−·2047	−·1758
Per cent of housing units owner-occupied	5	·0222	·0684	−·0684	−·1297
Average value of the housing-unit	6	·4105	·3557**	·2106	·1004
Average no. of rooms of owner-occupied units	7	2·1503	·2400*	−1·5553	−·1072
Average rent	9	·1519	·3792**	·2320	·3318**
Average no. of rooms of renter-occupied units	10	·0801	·0993	·4761	·3320**
Per cent of non-white	11	−·0105	−·0555	−·0188	−·0594
Per cent of 1·01 or more persons per room	12	−·2542	−·1853	−·0034	−·0016

** Significant at ·01 level.
* Significant at ·05 level.

To achieve a standard for comparison of five of the variables (6, 7, 9, 10, 12) a new analysis was run in which the beta weights were used. The results of this analysis indicated that none of the variables were significantly related to leukaemia incidence.

The multiple correlation coefficients of both estimating equations are significant at the ·05 level. The original nine variables have a multiple correlation coefficient of ·4984, while the standardised form of the data has a multiple correlation coefficient of ·4284. It may therefore be concluded that the analysis of the census tract data did not provide substantive results. It is apparent that the results at this scale do not explain the spatial pattern of leukaemia in Buffalo, New York. For this reason, the scale of analysis will now be changed to the block level to investigate whether a more detailed micro-level approach will produce more conclusive results.

THE BLOCK LEVEL

The block scale provides a more detailed spatial analysis of the relationships between housing and leukaemia. To allow comparison, however, the procedure used for the census tract scale will also be used in this instance. The only exception is the introduction of three new variables (4, 8, 13). They are: (4) per cent of housing units deteriorating; (8) per cent of housing units renter-occupied; and (13) persons per house.

Table 17.2 presents the regression and correlation analysis. Three of the variables are significant in both the simple and multiple equations. The first of these, (5) per cent of housing units owner-occupied, has correlation coefficients of ·1512 and ·1213 in the simple and the multiple correlation analysis respectively. The regression coefficient is positive, suggesting that

TABLE 17.2 *Block Scale Correlation and Regression Analysis*

Variables	Var. no.	Simple Regression		Multiple Regression	
		Regression coefficient (b)	Correlation coefficient (r)	Regression coefficient (b)	Maximum Partial correlation coefficient (r)
Per cent of housing units sound	2	·1189	·1355**	·0972	·0649
Per cent of housing units with plumbing	3	−·0307	−·0382	−·1949	−·1470**
Per cent of housing units deteriorating	4	·1205	·1307*	·0498	·0486
Per cent of housing units owner-occupied	5	·1781	·1512**	·2163	·1213*
Average value of the housing unit	6	−·0220	−·0325	·0992	·1141*
Average no. of rooms of owner-occupied units	7	−12·3658	−·4006**	−13·9873	−·3791**
Per cent of housing units renter-occupied	8	−·0504	−·0566	−·0966	−·1042*
Average rent	9	−·3370	−·2075**	−·0647	−·0544
Average no. of rooms of renter-occupied units	10	−12·7398	−·3463**	−6·2874	−·1809**
Per cent of non-white	11	·0118	·0064	·1187	·1043*
Per cent of 1 or more persons per room	12	−·2577	−·0574	−·4457	−·0868
Persons per house	13	−11·3445	−·1377**	·8525	·0103

** Significant at ·01 level.
* Significant at ·05 level.

increased rates of leukaemia are directly associated with increases in home ownership. However, the correlation coefficients are extremely low.

The second variable to be significantly related to leukaemia incidence is the average number of rooms in both (7) owner and (10) renter-occupied units. The negative regression coefficient is in direct contrast to the results obtained from the same two variables in the census tract scale, indicating that increased rates of leukaemia are associated with fewer room as in the census tract analysis. Thus, on the block scale the hypothesis that leukaemia is an infectious disease more likely to be transmitted in crowded conditions is supported. At the same time the negative regression coefficient associated with variable 13, persons per house, suggests that increased rates of leukaemia occur when there are fewer people in the home.

The importance of the remaining variables fluctuate, depending on the equation cited. For example, average rent (9) is significant at the ·01 level in the simple analysis but not significant at all in the multiple analysis. Too, differing from the census tract scale, the relationship is negative, i.e. increased rates of leukaemia are associated with decreased rates of average rent. It is important to note that the results from the block scale frequently are in direct contrast to those obtained from the census tract scale.

The results of the standardised variables indicate that average rent and average value of the home are not significantly related to the dependent variable. However, as in the original data, both average number of rooms in the owner and renter-occupied units are significantly related at the ·01 level.

Variables 12 and 13 which measure crowding in a home are both significant at the ·01 level. Variable 12 per cent 1·01 or more persons per room, exhibits a negative regression coefficient and an extremely low correlation coefficient. On the other hand, variable 13, persons per house, exhibits a positive relationship to the dependent variable. The multiple coefficient of correlation for the original data is ·5583 and for the standardised data ·5649. Both are significant at the ·01 level. It is apparent from the census tract and block scale that the selected standardised variables (6, 7, 9, 10, 12, 13) account for most of the explanation obtained by the estimating equation. Thus, the forthcoming analysis will utilize only six standardised variables.

Having evaluated the census tract and block scale, with no apparent substantive results, except for the average number of rooms in the owner and renter units, the analysis will therefore have to turn to still another scale represented by the tax districts.

AN AGGREGATED SCALE (TAX DISTRICTS)

At this point of the analysis the scale can be changed in two ways. Either one may analyse each diseased individual's home or one may aggregate the data. Both approaches will be used but the latter will be discussed first.

The data obtained for the individuals can easily be aggregated into fourteen tax districts shown in Fig. 17.1. The tax districts contain approximately five census tracts each. Again, all cases that occurred in a block within a particular district are aggregated and tested via regression correlation analysis.

The variables analysed are only the six that were standardised and that the previous analyses have indicated to be most important.

Table 17.3 gives the maximum partial correlation coefficients for the six variables of the fourteen tax districts. Of the fourteen areas, the multiple correlation coefficient is significant for seven of the areas. The significance level is ·01 for all seven. As shown in the table the correlation coefficients are high, ranging from ·9827 for district four to ·6914 for district ten. Of the variables tested, this indicates that they explain the incidence of leukaemia in those areas with some degree of reliability, even though the small number of observations entered into the analyses in some of the areas causes serious problems. Figure 17.2 shows the resulting high-risk areas within the city of Buffalo. One may note that there is almost a concentric zone around the core of spatially high risk areas. These may be considered low-risk areas.

More detailed examination of the variables indicate that variable 6, the average value of the home, shows no significant relationship in any of the fourteen districts. Variable 7, the average number of rooms in the owner-occupied unit, is significant in five of the seven spatially high-risk areas. The level of significance for each may be seen in Table 17.3. The regression analysis suggests that increased rates of leukaemia are associated with a decreased number of rooms in the owner-occupied units. This is very suggestive of a crowding condition and may support the theory of a viral aetiology.

Variable 9, average rent, is significantly negatively related to the incidence of leukaemia in four of the seven areas. However, in areas six and eight, the relationship is positive.

Variable 10, average number of rooms in a renter-occupied unit, is also significantly related to the dependent variable in four of the seven high-risk areas. The relationship is negative as for variable 7, average number of rooms in owner-occupied units. Again, increased rates of leukaemia are associated with decreased number of rooms in the renter-occupied units. This further supports the viral aetiology hypothesis.

Table 17.3 further specifies that variable 12, the per cent of 1·01 or more persons per room, is significant in only two of the spatially high-risk areas. Variable 13, average number of persons per house, is positively related to leukaemia with correlation coefficients equal to ·8215, ·7394 and ·6969 in the spatially high-risk area of two, four and eight. The relationship in this instance indicates that increased rates of leukaemia are associated with

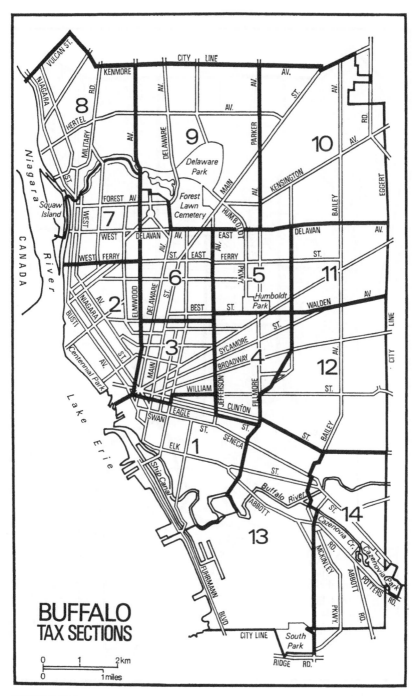

BUFFALO
TAX SECTIONS

FIGURE 17.1

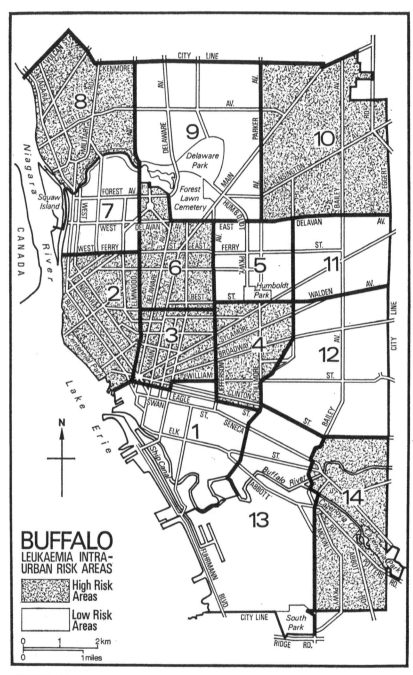

BUFFALO
LEUKAEMIA INTRA-
URBAN RISK AREAS

High Risk Areas

Low Risk Areas

0   1   2 km
0   1 miles

FIGURE 17.2

TABLE 17.3 *Tax Districts Maximum Partial Correlation Coefficients*

Variables	*Var. no.*				Fourteen Tax Districts			
		(1)	(2)	(3)	(4)	(5)	(6)	(7)
Average value of the housing unit	6	−·8743	·2960	−·1195	−·2219	·1054	·1982	·0088
Average no. of rooms of owner-occupied units	7	·4820	−·3308	−·4662*	−·1976	−·2344	−·5264**	·0298
Average rent	9	−·8629	−·4386*	−·3998	−·4797	·1900	·3992*	−·1169
Average no. of rooms of renter-occupied units	10	·8403	·8042**	·4524*	−·8543**	−·0767	−·3755	−·3049
Per cent of 1·01 or more persons per room	12	·7622	−·4354*	·1018	−·0448	−·1670	·3883*	−·1556
Persons per house	13	−·6584	·8215**	−·2184	·7394**	−·3454	·2503	·0565
Multiple Correlation Coefficients		·9450	·8691**	·8248**	·9827**	·6185	·8740**	·7164

		(8)	(9)	(10)	(11)	(12)	(13)	(14)
Average value of the housing unit	6	−·3824	·2866	·2374	·3058	·2839	−·8728	·0819
Average no. of rooms of owner-occupied units	7	−·7891**	·2149	−·2786*	·1422	−·7581**	−·8278	−·4167*
Average rent	9	·6065*	−·2072	−·2157	−·0067	−·3025	−·5734	−·0831
Average no. of rooms of renter-occupied units	10	−·5561*	·0444	−·1335	−·4091	−·1280	−·8636	−·3565
Per cent of 1·01 or more persons per room	12	−·3234	·2228	−·1870	−·2252	−·2919	−·6657	−·3017
Persons per house	13	·6969**	·0035	−·1462	·2252	·1962	−·6153	·2472
Multiple Correlation Coefficients		·9691**	·6403	·6904**	·7455	·9430	·9982	·7754**

** Significant at the ·01 level.
* Significant at the ·05 level.

more persons in the house. This is a definite condition related to crowding. Moreover, considering variables 7 and 10 (which are the average number of rooms in the owner- and renter-occupied units), in conjunction with variable 13, one may be further justified in suggesting that crowding is associated with leukaemia.

Of the three scales tested, the aggregation of the block data into tax districts seems most relevant; using this scale it was possible to identify spatially high- and low-risk areas. However, further associations of housing and leukaemia might possibly be found in the homes of diseased persons. A specific analysis of each diseased person's home may therefore result in other associations.

THE MICRO SCALE (THE HOUSING UNIT)

In an effort to determine if the characteristics of the house are associated with the diseased individuals, a random sample of homes that did not have leukaemia individuals will be selected. Each classification of the data is tested by application of the Kolmogorov–Smirnov test, which measures the maximum difference between two cumulative distributions.

TABLE 17.4 *Class of Dwellings. Smirnov Test*

Class of Dwelling	Leukaemia	Non-leukaemia	Difference
Hotel	·005	·000	·005
Duplex	·041	·011	·030
Apartment	·160	·028	·132*
Single	·576	·621	·039
Double	1·000	1·000	·000

* Maximum difference (Significant at the ·05 level.)

Table 17·4 gives the results for the type of dwellings. It may be noted that a maximum difference occurs in the apartment class. The maximum difference of ·132 is large enough to be significant at the ·05 level. This significance level indicates that there is indeed a difference in the type of dwelling inhabited by the leukaemia individuals. The difference is attributed to more leukaemia victims living in apartments than in other types of dwellings.

A further nine groups of house construction characteristics were tested in a similar manner. The characteristic and classifications used are given in Table 17.5, but for none of the categories employed is there a difference between the housing units of the leukaemia and non-leukaemia individuals.

However, two additional variables are tested for variations via a difference of proportions test. These variables are the existence of fireplaces and laundry facilities in the home. The frequency of laundry facilities in the homes with leukaemia is ·6141 and the frequency of laundry facilities in the homes without leukaemia is ·6081. The difference of proportion test yields a result of ·157 which, when entered into a normal curve, does not indicate a significant difference between the two distributions.

The frequency of fireplaces was tested by the same method. In the homes in which leukaemia occurred, the fireplace frequency is ·42847 and the

TABLE 17·5 *House Construction Characteristics*

Type of foundation:	brick; posts; concrete; stone.
Type of basement:	no basement; $\frac{1}{4}$ basement; $\frac{1}{2}$ basement; full basement.
Construction type:	tile; concrete; stone; brick; frame.
Exterior construction:	paper; pressed brick; stucco; combination brick; shingles; siding.
Type of roof material:	board; iron; asbestos; tile; gravel; slate; prepared shingle; shingle.
Roof structure:	gambrel; dormers; flat; hip; gable.
Type of outside trim:	galv. iron; terra cotta; concrete; stone; wood.
Type of heating:	central station; steam; hot water; stove; furnace.
Source of household power:	electric; gas and electric.
Condition of housing unit:	Poor; fair; good.

homes of the non-leukaemia individuals had a frequency of ·32012. The difference of proportions test yielded a result of 2·923, which gives a significant difference. In a two-tailed test the probability of obtaining a value as large or larger than 2·923 is ·0032. This is significant at the ·01 level, which means that there is a difference between the two distributions.

Finally, the average number of rooms were found to be quite different between the two samples. Thus, the average number of living-rooms in the leukaemia homes was 2·5 as compared with 3·3 in the sample of non-diseased. Also, the average number of bedrooms is considerably lower; the diseased individuals had an average of 2·2 bedrooms per house. Overall, the average number of rooms is 4·7 in the diseased sample as compared with 6·2 in the non-diseased sample. The results suggest a greater possibility of crowding in the homes of the individuals who have leukaemia.

I

SUMMARY

This paper has presented material on housing variables for four different scales. There were only a few of the variables and only one scale which proved to provide substantive results.

The census tract scale suggested possible relationships but the correlations were very weak. Average rent, average value of the home and average number of rooms in the owner- and renter-occupied units were significant at the ·01 level. However, the correlation coefficients were very low and none exceeded ·4500.

The block scale produced similarly poor results. Thus, there were significant correlations of several variables, but the correlation coefficients were well below accepted standards. Interestingly, the average number of rooms in the owner- and renter-occupied units were significant as in the previous census tract analysis, but the direction of the relationship had changed from positive to negative. The standardised variables for the block scale also proved to be significant but the correlations were low. For the census tract scale the relationship of the number of rooms was positive, while for the block scale the reverse was the case. Comparing the two scales, it may be concluded that both fail to estimate the incidence of leukaemia with any reliability.

In contrast to the census tract and block analysis, the results obtained from the aggregation of block data into districts were remarkable. In particular, it was possible to identify spatially high- and low-risk areas. The multiple correlation coefficients are high, varying from ·6904 to ·9827. The variables tested were identical to the census tract and block scale. The variables which are significant most times were average number of rooms in the owner- and renter-occupied units. The relationship was usually negative, i.e. increased leukaemia rates were associated with fewer rooms in the unit. However, the relationship was reversed in two instances.

Characteristics of individual houses provided the data for a specific micro-analysis. Several aspects were tested but failure to reject the null hypothesis prevailed. Only the presence of fireplaces in the house and the number of rooms in the home were found to be significantly different between houses with and without leukaemia victims. The second aspect of those variables supports the results obtained from the aggregated data. Indeed, increased rates of leukaemia exist where there are fewer rooms in the home. As to the fireplaces, it is possible the socio-economic status is higher in the individuals who get leukaemia and this may reflect the presence of fireplaces in the leukaemia homes. If this were the case, however, one would expect more rooms in a home and probably less crowding. On the other hand, a fireplace is in fact an open hole in the home and it is a possible route for various vectors. Also burning of certain products in the fireplace may produce by-products which could cause leukaemia. Obviously, the presence of fireplaces is open to wide conjecture.

The significant finding of this paper is the identification of spatially high- and low-risk areas in the intra-urban setting of Buffalo. It is evident that causation has not been established but association was definitely determined.

REFERENCES

1 MCPHEDRAN, P. and HEATH, C. W. (1969) *Cancer*, **29**, 2021–5.
2 GILMORE, H. R. and ZELESNICK, G. (1962) *Pennsylvania med. J.* **65**, 1047–1049.

18 Food Contaminants and Oesophageal Cancer

N. D. McGlashan

Cancer of the oesophagus (gullet) differs widely in its incidence in different localities.[1] Although geographical comparisons are made difficult by variations in the nature of statistical data available in different countries, this disease may be at its commonest in certain less developed countries with poorer records. Even in developed countries the differences in incidence are striking; in 1962 standardised oesophageal cancer mortality rates for males varied from 2·8 per 100,000 (Australia) to 11·2 (France) and amongst females, from 0·9 (United States, whites) to 5·5 (Finland).[2] Amongst African countries, such accurate figures are not obtainable, but certain measures can be used from comparison. Oettlé summarised the known state of knowledge in 1962[3] and showed in cartographic form the marked 'patchiness' of occurrence of oesophageal cancer even within sub-Saharan Africa. At the same time he and others reported a marked rise in case numbers in the two decades after the war.[4] This, he felt, was markedly more than could be accounted for by better medical or diagnostic facilities or by the increasing willingness of Africans generally to seek European-type hospital treatment.

To the geographer Oettlé's maps were particularly interesting since the distributions could not be shown to coincide with any known factor of the physical environment. Areas of high incidence seemed to vary widely in their physical geographical characters. For example Capetown, Bulawayo, Nairobi and Mombasa fall into Oettlé's 'numerous' frequency class. Also, on his map, 'not seen' reports from hospitals are spatially interspersed between the 'numerous' reports.

Factors in the social environment and those associated with a local custom seemed to provide more promising lines of investigation and here certain correlations were recognised. Oettlé stated 'cancer of the oesophagus is well known to be associated with alcoholism'[2] and, in 1964,[3] he quotes 'a strong correlation with smoking'. He also mentions the *possibility* of certain metallic elements, lead, brass and bronze (but not zinc) being cancer-causing agents.

Amongst the studies undertaken in Africa before 1965, one of the most interesting was that of Burrell in East London.[5] On a large-scale map of the African locations he showed that the homes of oesophageal cancer patients, predominantly males, clustered around *shebeens* (illicit spirituous liquor stores). This pattern altered mysteriously immediately after police raids closed those sources of alcohol, and eventually a new, clustered distribution of disease could be recognised.

DISEASE DATA IN CENTRAL AFRICA

Cancer, an uncontrolled (malignant) division of the replicative cells of the body, usually affects primarily one particular site within the human body. If the condition is untreated secondary tumours will occur affecting other organs often far removed from the original site. 'In Africa oesophageal tumours occur most frequently in the middle and lower parts of the oesophagus.'[6] Whilst radiological examination will rarely fail to diagnose malignancy, clinical diagnosis alone may not be reliable,[2,6] even in areas of frequent occurrence of cases and by those with clinical experience of the disease.

The fact that the prognosis for oesophageal cancer patients is not good may have a bearing on case reporting, as the Bantu are often fatalistic when they recognise in themselves symptoms which they believe to be incurable. Such persons may deliberately choose to die at home untroubled by the palliative efforts of medicine.

Early in 1965 a geographical pathology survey was undertaken in central Africa from the University of Zambia which aimed to collect data concerning fifty-five different disease conditions and a number of previously unpublished environment factors. These factors were selected on the criterion of having, or being thought to have, an aetiological relationship with one or more of the diseases being studied. Oesophageal cancer, as well as five other forms of cancer, was included and questions concerning beverage and spirituous liquor usage and form of tobacco use locally preferred were also asked.

In all cases the sources of information were the hospitals, which therefore formed a network of stations conforming approximately with the population distribution.[7] A team of a geographer and a nursing sister visited each of 103 hospitals on an 8,000 mile road tour, passing through four countries, of which two, Zambia and Malawi, were totally covered by the survey. Oesophageal cancer patient numbers (in absolute terms) were obtained only by doctors' impressions by recall of cases in answer to the question, 'How many cases are seen in this hospital in an average year?' Answers to such a question will be more correct in cases of a rare or dramatic medical condition, but these reports are clearly subject to error and must be interpreted with that caution. Social customs in the area immediately served by each hospital were explored by questioning senior local African members of hospital staff.

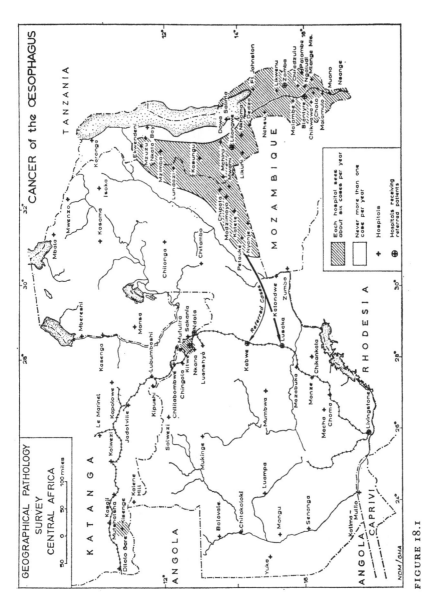

FIGURE 18.1

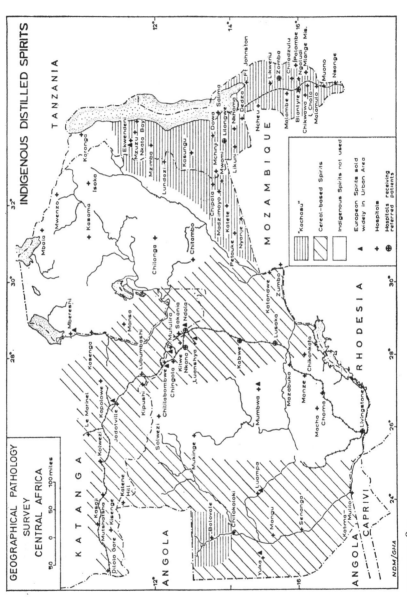

FIGURE 18.2

The maps illustrate cartographic generalisations of the individual hospitals' reports of oesophageal cancer case numbers (Fig. 18.1) and local alcoholic spirit-taking customs (Fig. 18.2). In fact, even during the journey, the correlation began to emerge and some first samples of local spirituous liquor were collected for later analysis. With all reports available, a chi-squared table was drawn up to test the significance of the spatial association. In Table 18.1 the numbers in the cells represent reporting hospitals and the null hypothesis is that there is no spatial correlation between the disease and the drinking of this particular form of alcohol. The null hypothesis is rejected at $p < 0.001$ level. Similar tables were drawn up by

TABLE 18.1 *Reporting Hospitals*

Numbers of Oesophageal Cancer Cases Annually		Nil	<2	2−11	12−51	Total	
Kachasu drunk locally	Yes	8 (22)	10 (27)	17 (46)	2 (5)	37	Observed
	No	38 (60)	21 (33)	3 (5)	1 (2)	63	
Kachasu drunk locally	Yes	17·02	11·47	8·51		37	Expected
	No	28·98	19·53	14·49		63	
All hospitals		46	31	23		100	

$$n = 2 \qquad \chi^2 = 28 \qquad p < 0.001$$

(a) Numbers in brackets indicate the percentage in that morbidity column of all hospitals in that drinking category.
(b) Of 103 hospitals visited, three made no report on alcoholic spirit use and are omitted from this table.

computer programme for each disease against each socio-environmental factor in pairs.[8] The oesophageal cancer map showed inverse distributional similarity with use of alcoholic beverages based on millet ($p < 0.005$). No oesophageal cancer cases were reported where cassava is the stable food ($p < 0.001$), but the disease is commoner where maize is the staple foodstuff ($p < 0.001$).

A spatial correlation of high significance which could be easily tested was that with alcoholic spirits, and so this parameter was specifically selected for more detailed study. *Kachasu*, also known as *Malawi Gin*, is distilled from a preparatory beverage of maize, maize-cobs and sugar in a variety of types of crude equipment.[9] Metallic containers are used in combination with tubes, including bicycle frames and discarded car exhaust systems, ingeniously brought into service as home-made condensers. A profitable but, in Zambia itself, illegal trade is carried on in areas remote from police surveillance.

Kachasu is a word which may be of Portuguese origin since a distillate of a similar name is used in Brazil. This suggests a diffusion of the practice of alcoholic distillation following Portuguese advances into south-central Africa via the Zambesi and Shiré river valleys. In this connection, is it mere coincidence that the only other area in Zambia where *kachasu*-type alcohol is prepared is on the Angola (Portuguese) border in the far west? Another example of possible cultural innovation was found in the Transkei of South Africa. Here a type of utensil radically different from those of central Africa is used to prepare alcoholic spirits. Here the beverage is heated in a large clay pot sealed by a metal bowl of cold water. The underneath surface of this bowl acts as a condensing surface and the condensate falls back into a smaller receptacle suspended inside. Identical apparatus was later examined in south India's prohibition states, and it is tempting to speculate that the method was taken to Africa by the indentured sugar plantation workers from India arriving in Natal in the 1890's. The bowls which provide the condensing surface, both in India and Africa, have a brief useful life since they are corroded into holes by the distillate within two or three months.

CHEMICAL ANALYSES

Early results of analytic work upon *kachasu* showed that better quality spirits may have up to 30 per cent ethanol but that all characteristics of strength, flavour and degree and type of contamination vary within very wide ranges. This is not surprising in view of the variety of preparative procedures and the secrecy needed to avoid detection. Many of the samples of spirits had been contaminated with zinc in amounts varying up to 65 mg. per litre and/or copper up to 58 mg. per litre. The source of this zinc lies probably in the galvanised iron drums used as fermentation vessels. The copper is ascribed largely to contamination with copper-rich soils and, to a lesser extent and only in certain cases, to the use of copper tubing in the distilling apparatus.[10]

Since it is probable that metallic containers and tubes were far less widely available in African rural areas in the past, change of custom with regard to preferred apparatus may be related in time to the reported increase in oesophageal cancer over the last twenty-five years or so. It would also seem likely that, following African social custom, men are greater drinkers than their womenfolk, thus accounting for their higher ratio as cancer patients. Information on individual drinking habits and sex differences in these habits was not asked for because both preparation and imbibing were illegal in Zambia and Malawi. One study in Durban, however, has suggested a positive correlation amongst hospital patients between oesophageal cancer and spirituous 'concoction' drinking.[11] Even if today's habits were fully known, interest should focus upon those prevailing at a date earlier than today by the latent period of this cancer.

Attention next centred on those parts of Zambia and her neighbours where oesophageal cancer was rare or unknown in order to analyse their alcoholic preparations and to see whether their chemical composition was similar or different. In areas (such as north-west Zambia) where spirits other than *kachasu* are prepared, those alcoholic drinks sampled were found to be similarly contaminated with zinc or copper or both. The only areas without contamination were those where distilling is not customarily carried on at all, such as the northern plateau of Zambia towards the Tanzania border. Honey beer, for instance, seems not to acquire such a metallic load in its preparation. This may be, in part, because it is not heated but merely warmed gently beside a fire during fermentation.

Short visits to India[12] and South Africa[13] by the author now permitted collection of a variety of local alcoholic spirits from these countries, where quite different methods and ingredients are used. In both these countries methylated spirits and/or varnish are added to illicit drinks and, in Natal, dead rats and other animals are added to give 'body' to the drink. This latter practice is also authoritatively reported from French-speaking parts of Canada.[14] Meanwhile from the laboratories of the British Manufacturing Industries Research Association at Leatherhead came the first report of the detection of small amounts of dimethyl-N-nitrosamine (DMN) in earlier Zambian samples.[15] Eight samples of liquors distilled from fermentations of maize husks and sugar gave indications of nitrosamine at concentrations of one to three parts per million – a level likely to prove carcinogenic in man.

Dimethyl-N-nitrosamine is a chemical compound of the N-nitrosamine group, many of which are known to cause cancer of certain specific sites of the body, in particular, in experimental animals, the liver and oesophagus.[16,17] DMN had been detected in fish-meal and cattle-cake[18] but had not previously been reported in any human foodstuff.

Geographical interest now centred on the question of whether there was a definable boundary to a region of contaminated alcohol, or whether N-nitrosamines occur world-over in home-brewed alcohol. Biochemical interest centred on the source of DMN in alcohol samples and the mode of its occurrence. It was expected that one discipline's answer, whichever was first arrived at, might assist in solving the second query.

One major problem, unsolved to date (October, 1970), arose in the analytic work. Detection of N-nitrosamines in low concentrations is extremely difficult and adequate procedures are not available to concentrate these compounds to enable positive and unequivocal identification.* Results

---

* Recent work (Oct. 1971) suggests that the contaminant may be N-nitrodimethylamine, also a potently carcinogenic substance.

obtained by polarography or by gas chromatography or by irradiation with ultra-violet light (when N-nitrosamines should release measurable nitrite) are cumulatively highly suggestive, even indicative, but not concrete proof. Positive biochemical identification would require large amounts of N-nitrosamines concentrated by gas chromatograph and passed through a mass spectrometer.

On the other hand, those responsible for public health, necessarily cautious, may feel obliged to accept *indications* of the presence of a potent known carcinogen even when it is merely suspected. When, as in this case, a long history of epidemiological evidence is adduced, the case may seem compelling.

OTHER GEOGRAPHICAL AREAS

With the help of a Medical Research Council grant (No. G-969/63/B) collections of samples were obtained from further afield: Ghana, Nigeria, Senegal, Kenya, Uganda, Eire, Canada and Yugoslavia. Wherever the distillation was illegal, necessitating crude methods and some degree of secrecy as in all these countries, the analyses indicated the presence of DMN. Furthermore, the presence of DMN was indicated whether the drink was a distilled spirit or merely a fermented beer. The explanation put forward is that N-nitrosamines would distil over with the spirit under most local and uncontrolled conditions where deliberate procedures to prevent this would be absent. Such a procedure might mean some loss of nitroso-contamination with resulting lower values of DMN in the spirit than in the original beverage. In general this result was found to be so. In view of the variety of climates sampled and the different bases of the fermented beverages used in these widely separated sources of the samples, it seems safe to assume that this type of contamination is not limited to certain local areas but that it is a truly world-wide phenomenon. By contrast, however, commercially bottled Scotch whisky, exhaustively tested, was found not to be contaminated with N-nitrosamine compounds.[19] Gas chromatograph linked to mass spectrometer confirmed, in the case of whisky, what had been established only by polarography for certain British gins, Malawi Gin, Uganda Warigi and Iranian Ettehadiah Arak, namely that commercial bottled alcoholic spirits seemed generally free of DMN.

An explanation is required of the observation that N-nitroso-contamination is not found in commercially bottled spirits. Two possible procedures would remove N-nitrosamines if these were present. Activated charcoal filtration is employed upon certain colourless liquors and fractional distillation is widespread in commercial spirituous liquor production. Neither procedure has been reported in connection with any of the home-brew preparations sampled and found to be contaminated. As an attempt to suggest a cleansing method practicable at village level, Zambian *kachasu* was filtered through ordinary village woodburner's charcoal and recovered through school quality blotting paper. This spirit was shown (admitting the limitations inherent in analytic procedures) to have lesser amounts of N-nitroso-contamination. Expert indigenous *kachasu* testers were unable to detect an alteration in their drink, but the procedure is messy, time consuming and unlikely to have wide appeal.

OCCURRENCE OF N-NITROSAMINES

To consider the biochemical mode of occurrence of DMN, a renewed visit was paid to Kenya where the highest concentrations of DMN in any samples had been found. Samples were taken of all the separate ingredients of several different types of alcoholic preparation and then further samples were taken, at each stage, as these ingredients were added to the mixture

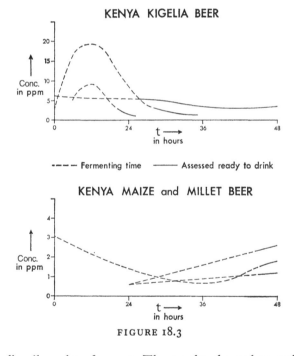

FIGURE 18.3

and eventually allowed to ferment. The results showed no polarographic activity in the ingredients sampled (at the peak potential ascribed to N-nitrosamines) until fermentation had set in. Then the general pattern is a sharp rise in DMN, as estimated by polarography, followed by a decrease again. The decrease, however, never falls away through time to nil. The general shape of the curves is as shown in Fig. 18.3 where values of time, *t*, vary from thirty-six hours to one week, depending upon the type of beverage and where concentration is only relative because values obtained from polarographic peaks are suspect in absolute terms. The contaminant amount, in any case, varies between different breweries and even between different brews.

The interpretation of the curve is uncertain. There is some evidence that DMN may be formed by wild yeasts acting in the nitrogen-rich fermenting beverage, regardless of the carbohydrate source. The decline from the highest values may be related to changing pH in the liquor during fermentation.[20]

NORTHERN IRAN

Finally a brief visit was paid to the Mazandaran province of Northern Iran where high oesophageal cancer incidence, particularly amongst women,[21] is being investigated in a continuing project by the Institute of Public Health Research of the University of Teheran. In this devout Muslim area, alcohol intake is probably unimportant. The results described above had implied that *fermentation* might be the key to DMN production and then alcohol would be only one amongst many possible vehicles for the carcinogenic contaminant. In Iran grape vinegar is prepared by a fermentation process and garlic and brinjal pickled in this vinegar were found to be items of local diet which were contaminated with DMN. Whether a sex difference of vinegar consumption occurs is not known, but it was desired to uncover an item specific to female diet to attempt to account for their high oesophageal cancer rates. In Mazandaran, Mahboubi had drawn attention to deviant dietary behaviour by women, especially of the Turkoman tribe, whose territory extends into neighbouring parts of the U.S.S.R., where high cancer rates are also reported. During pregnancy and as a general antidote to digestive discomfort, these women eat a mixture known as *anarteen*. This is prepared from raisins crushed by pestle and mortar together with pomegranate seeds and black peppercorns. The resultant sticky mixture is eaten raw and eaten consistently over many child-bearing years. A correlation with parity and/or age of women has not yet been investigated. Three separate samples of *anarteen* showed contamination with dimethyl-N-nitrosamine.[22]

CONCLUSION

It may seem that the dietary customs of Iranian females are a far cry from the *shebeens* of Transkei or the village carousals of eastern Zambia. The original and basic clue to this line of reasoning was the recognition in central Africa of two geographical patterns of distribution with significant similarity. The line of evidence leads, somewhat indistinctly in places, to another apparently different but actually related spatial correlation of a contaminated female dietary constituent in a locality of high oesophageal cancer incidence.

REFERENCES

1 DOLL, R. (1967) 'Prevention of cancer: pointers from epidemiology'. Nuffield Provincial Hospitals Trust, Oxford.
2 OETTLÉ, A. G. (1962) *S. Afr. Cancer Bulletin,* **6,** 112.
3 OETTLÉ, A. G. (1964) *J. natn. Cancer Inst.* **33,** 383.
4 ROSE, E. F. (1965) *S. Afr. med. J.,* **39,** 1098.
5 BURRELL, R. J. W. (1957) *S. Afr. med. J.,* **31,** 401.
6 PALMER, P. E. S. (1967) Natn. Cancer Inst. Monograph No. 25.
7 MCGLASHAN, N. D. (1968) *Med. J. Zambia,* **2,** 17.
8 MCGLASHAN, N. D. and BOND, D. H. (1970) *Can. Geogr.,* **14,** 243.
9 MCGLASHAN, N. D. (1969) *Gut,* **10,** 643.
10 REILLY, C. and MCGLASHAN, N. D. (1969) *S. Afr. J. med. Sci.,* **34,** 43.
11 BRADSHAW, EVELYN and SCHONLAND, MARY (1969) *Br. J. Cancer,* **23,** 275.
12 MCGLASHAN, N. D. (1969) *Ind. med. J.,* **63,** 169.
13 MCGLASHAN, N. D. and WALTERS, C. L. (1969) *S. Afr. med. J.,* **43,** 800.
14 Royal Canadian Mounted Police, Montreal (1970) pers. comm.
15 MCGLASHAN, N. D., WALTERS, C. L. and MCLEAN, A. E. M. (1968) *Lancet* ii, 1017.
16 MAGEE, P. N. and BARNES, J. M. (1956) *Br. J. Cancer,* **10,** 114.
17 MAGEE, P. N. and BARNES, J. M. (1967) *Adv. Cancer Res.,* **10,** 163.
18 *Lancet* (1968) ii, 107 (editorial).
19 MCGLASHAN, N. D., PATTERSON, R. L. S. and WILLIAMS, A. A. (1970) *Lancet* ii, 1138.
20 REILLY, C. (1970) pers. comm.
21 MAHBOUBI, E. (1970) pers. comm.
22 MCGLASHAN, N. D. (1970) *Int. Pathology,* **11,** 50–53.

Part V
Disease Diffusion

19 River Blindness in Nangodi, Northern Ghana: A Hypothesis of Cyclical Advance and Retreat[*]

J. M. Hunter

It has recently been estimated that the incidence of blindness in northern Ghana is 3,000 per 100,000, as compared with about 200 in Europe. A little less than half of the cases in northern Ghana are the result of onchocercal infection, and where the local incidence of blindness is 10 per cent or more, nine out of ten cases are due to onchocerciasis,[1] or 'river blindness'. Onchocerciasis is endemic in Mexico, Guatemala and Venezuela; it is believed to have been carried to the Western Hemisphere by African slaves.[2] In Africa it is endemic principally in the savanna belt south of the Sahara, from Senegal in the west to Uganda and Kenya in the east.

PATHOGENESIS

River blindness is caused by a parasitic worm of the filaria group, *Onchocerca volvulus*. When an infected human subject is bitten by the female of the fly vector (*Simulium damnosum* in Ghana; other species elsewhere), microfilariae are sucked through the proboscis into the fly's stomach. Within twenty-four hours they migrate into the thorax muscles, where they undergo a change without which they could never grow into adult worms. At the end of a week, in the infective larvae stage, they move to the salivary glands, from which they may be implanted in the skin of another human subject when the next blood meal is taken.

Adult filariae breed, in the human subject, in an interlocked mass around which a reactive fibrosis develops, forming a subcutaneous nodule (onchocercomata) ranging from the size of a pea to that of a golf ball. It is probable that there are also free adult filariae in the skin tissues outside the nodules. Microfilariae are hatched out by ovoviviparous females and spread freely

[*] Reprinted from *The Geographical Review*, (1966), **56**, 398–416. The author is deeply grateful to his friends and interpreters, Anaara Anafu, Thomas A. Apasnaba, and Nyaba N. Yembilah, for their indefatigable support during the survey, and to the late Nangodi-Naba and his chiefs and elders for their kind cooperation.

from breeding sites through the subject's subcutaneous tissues, where they remain unchanged throughout their life until picked up by a fly. Female adult filariae attain a length of 33·5–50 centimetres and a transverse diameter of 270–400 microns; males, 1·9–4·2 centimetres and 130–210 microns. Microfilariae vary greatly in size, the smaller forms measuring 150–250 microns in length, the larger forms 285–360 microns. The life of the adult worm is 10–15 years and that of the microfilaria 3–9 months. As many as 300 microfilariae may be found in a single skin biopsy one millimeter in diameter. The number in a heavily infected subject runs into billions.[3]

By preference, the fly in West Africa bites the lower limbs (88 per cent of all bites in one survey, of which 76 per cent were on the calves and shins[4]), and normally the infection spreads upward, sometimes taking years to reach the eyes. Besides the nodules, a heavily infested skin will show signs of atrophy, diminished elasticity, and the thickening known as pachydermia, or 'elephant skin'. In time the growing microfilarial population invades the eyeball, through sheaths of perforating blood vessels, and accumulations of dead parasites set up inflammatory reactions in the ocular tissues that lead finally to blinding lesions.

In northern Ghana, where the rate of infectivity is exceptionally high, some degree of natural immunity has developed, and the invasion of the ocular tissues does not necessarily result in blinding lesions. Only a massive invasion can break through this immunity.[5] Of considerable importance, too, is the fact that dietary balance or imbalance can respectively retard or accelerate the development of onchocercal lesions. For example, in the rain forests of Onitsha Province, Nigeria, where the intake of vitamin A is very high owing to a diet rich in palm oil, there is no ocular onchocerciasis despite an infectivity rate of 75 per cent.[6] On the other hand, inhabitants of the savanna, with only about one-quarter of the Onitsha vitamin A intake, enjoy no such dietary protection.

Clinical treatment, yet to be practised on an effective scale in West Africa, may include excision of nodules and the administration of vitamins and filaricidal drugs. Difficulty has been experienced in finding a satisfactory drug for adult filariae, but one drug, Hetrazan, is a successful microfilaricide.[7] A subject's reaction to the toxic and other effects of certain drugs may be severe. To the writer's own knowledge, many Africans prefer to suffer impaired and deteriorating vision rather than undergo a course of filaricides. New drugs will of course eventually overcome this antipathy, which is based on earlier and perhaps premature application of less efficient drugs. In the general context of economic development, attention may be drawn to the high capital costs of provision of adequate medical services; and in view of the impoverished agricultural resource base of most of the African savanna, it is hardly surprising that vast areas enjoy as yet only token, or at best inadequate and overworked, medical services. Nangodi, the area discussed here, is no exception.

ENTOMOLOGY

The main vector of onchocerciasis in West Africa, *Simulium damnosum* (the common name of the genus is 'black fly'), is catholic in its choice of host, feeding on cows, donkeys, sheep, goats, dogs, birds, game and man. Fully grown, it is less than one-quarter of the size of a house fly. After a blood meal the female lays her eggs in swiftly flowing water ($3\frac{1}{2}$–$4\frac{1}{2}$ feet a second) on substrates such as rocks and suspended vegetation at a depth of 0–6 inches. The larvae measure about a quarter of an inch and are attached to the substrates by a ring of hooks. The oxygen requirements of the larvae are very high, hence the restriction of breeding sites to swift and, ideally, bubbling (that is, oxygenated) streams. After a gestation period of 7 days, about 250 eggs are laid in each oviposition. These hatch in 36 hours. Larval life lasts 10 to 13 days, and pupal life about 4 days, after which the new flies emerge. After oviposition the fly takes another blood meal in order to begin a further reproductive cycle.[8]

Flies are not born parasitised, and obviously not every fly becomes infected with microfilariae. Infection rates in a given area vary from day to day. One survey showed variations from 8 per cent to 51 per cent within a single month.[9] Cattle and game serve as reservoirs of onchocerciasis, and it is believed that the parasites they carry can be transmitted to man, though this has not yet been experimentally demonstrated. Since breeding grounds are necessarily restricted to certain riverine areas, the flight range of the fly is of some importance. Twelve miles seems to be the most likely maximum, but further transmission by gusts of wind must also be a consideration. *Onchocerca volvulus* has a deleterious effect on the flight muscles, and the flight range of parasitised flies is therefore less than that of uninfected flies. This has the general effect of restricting heavy infectivity to the principal breeding sites. Operating to the same end is an apparent preference of flies that have begun breeding (parous flies) to keep close to breeding sites, as compared with young (nulliparous) flies, which disperse more widely. Evidence of black-fly behavior in Canada may be relevant on this point (Fig. 19.1).[10] Parous flies that have taken blood meals are, potentially at least, vectors of onchocerciasis; hence their tendency to concentrate around breeding sites has the effect of limiting the spread of infection to areas smaller than the flight range of the fly would suggest.

In Ghana and Upper Volta onchocerciasis is concentrated between latitudes 10°30'N and 12°N. Beyond 12°N little onchocerciasis occurs, even where the vector is abundant. The reasons are not clear. The shorter rainy season has been suggested as a factor; possibly also the fly may be zoophilic in those areas rather than anthropophilic; or it may be that the disease is still in process of introduction.

An important gap in the entomologist's knowledge of the life cycle of the fly is how it survives the dry season. For six months or so, when the rivers are not flowing, the fly disappears without trace; yet within hours of resumption of stream flow after the first showers of the rainy season it reappears in adult form. Aestivation has been postulated,[11] but so far no

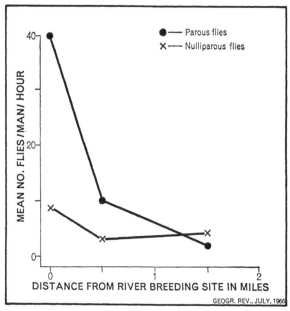

FIGURE 19.1 Numbers of nulliparous and parous *Prosimulium mixtum* at three sites near Ottawa, Canada, May 9 to June 8, 1960. Parous flies have completed an ovarian cycle, nulliparous flies have not.
(From: Davies, L. (1961) *Canadian Entomologist*, **93**, 1125.)

supporting evidence has been produced. Repeated and careful examination of dry-season river beds and riverine vegetation has revealed various life forms, but no *Simulium*.

It is the writer's view that the flies, or at least a sufficient number of them to initiate a breeding cycle, move northward in the tropical maritime (T_m) air mass that advances seasonally from the Gulf of Guinea, thrusting wedge-like under the tropical continental (T_c) air of Saharan origin. During each rainy season part of the fly population of permanent breeding sites in the more humid south would thus be able to migrate northward. This essentially geographical idea is now under investigation by a resident entomological research unit in northern Ghana, and it appears likely to bear fruit.[12]

Ideally a fully coordinated attack on onchocerciasis would include, on the medical side, minor surgery and the distribution of filaricidal drugs and vitamins, together with radical dietary readjustments in the worst-afflicted areas, all aimed at reducing and eventually eliminating the human reservoir of infectivity. On the entomological side, regular treatment of breeding sites, permanent and ephemeral, large and small, would certainly reduce the vector population. Treatment of the main rivers with larvicides has been proposed, and, indeed, successfully demonstrated in many areas (including Kenya[13]), but the problem of reinfestation from tributary streams and ephemeral torrents remains. The medical programme appears to offer better prospects in the long run, but either course, if pursued on a full scale, would seriously strain the financial and professional manpower resources of those developing countries where the disease is prevalent.

SURVEYS IN GHANA

Survey work in Ghana began in 1945 when Ridley made a limited but intensive ophthalmic study, in which fifty-one cases of ocular onchocerciasis were diagnosed in a single village.[14] Waddy, a Medical Officer of Health, followed with a reconnaissance survey of a wide area of the north in 1948–9.[15] His graphic descriptions rightly insisted on the importance of a disease that had thitherto been largely overlooked. In 1952 the Royal Commonwealth Society for the Blind sponsored both entomological and ophthalmic surveys, directed respectively by Crisp and Rodger.[16] At the same time a permanent central-government system of Medical Field Units was established, under Waddy's control, charged with the enormous task of ascertaining the distribution and intensity of, and attempting to control, major endemic diseases, of which onchocerciasis was but one.

The picture that emerges from the medical surveys is that onchocerciasis is endemic in riverine areas in the Guinea savanna woodland of northern Ghana. The incidence is higher on granitic and metamorphic rocks than on sandstones, where gentler stream profiles provide fewer suitable breeding sites. North of 10°30′N, along the Black Volta, the Kulpawn, the Sissili, the Red Volta, and the White Volta (Fig. 19.2), the disease is hyperendemic. Representative of this upper zone of hyperendemicity is Nangodi, a small traditional state about sixty square miles in area on the right bank of the Red Volta bordering on Upper Volta.

THE RETREATING FRONTIER OF SETTLEMENT IN NANGODI

The traditional politico-military state of Nangodi, ruled by the Nangodi-Naba, falls into two distinct territorial divisions.[17] In the west nearly 14,000 people, arranged in 28 sections and 10 chiefdoms, occupy an area

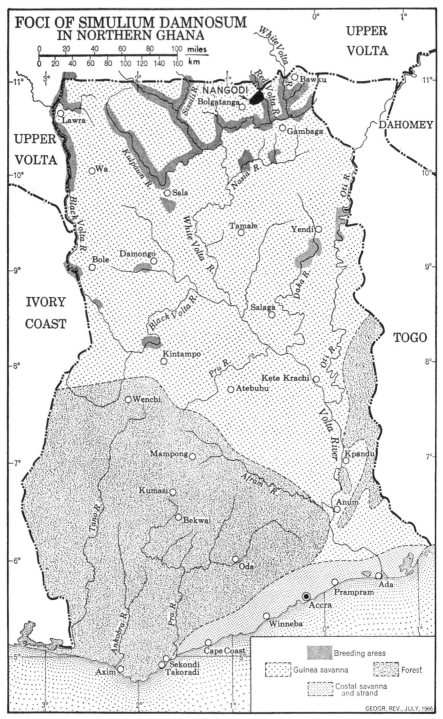

FIGURE 19.2 Principal foci of *Simulium damnosum* in northern Ghana. Breeding areas are after Crisp, G. (1956) who notes that the distribution of the fly is certainly more extensive than shown, since he has excluded doubtful areas.

of 37·2 square miles. In the east, toward the Red Volta, 22·6 square miles – some 38 per cent of the total area of Nangodi – are without a single house or farm. This area, once intensively farmed, is now completely abandoned because of river blindness.

The line between the two regions (Fig. 19.3) indicates what we may call the frontier of settlement, or, alternatively, the frontier of blindness. It is not static or in equilibrium. There is evidence that settlement has been

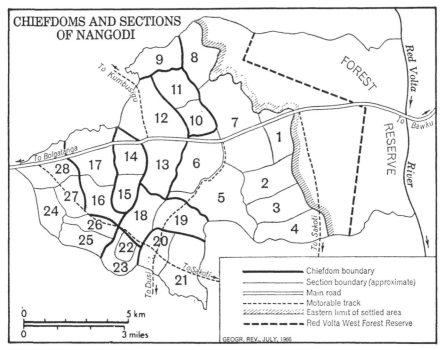

FIGURE 19.3 Chiefdoms and sections are the basic politico-military divisions of the Nagodi traditional area. Names and populations are given in Table 19.1. With one exception (Zoagin, Section 26, which was surveyed by chain and compass in connection with the study of settlements), boundaries are approximate, sketched by the author during a ground survey in 1963, using air photography of 1960. The eastern limit of the settled area was also plotted from the 1960 photography. The part of the Red Volta West Forest Reserve was shown demarcated in 1941 to include an already abandoned area.

retreating, or blindness encroaching, for several decades, perhaps half a century, and that the retreat continues today. Except for Kalini (Section 1), which is the Nangodi-Naba's section and therefore more resistant to population decrease because of its institutional functions and status, and Nakpaliga (Section 7), which lies mostly behind Kalini, all sections along the frontier are greatly shrunken in population, especially as compared with the sections in the west (Table 19.1). Population trends need not be described in detail here, but mention may be made of the general decline of the frontier sections. Two examples will suffice. During the intercensal period 1948–60, the number of houses in Gongo (Section 8) decreased

TABLE 19.1 *Estimated Population of Chiefdoms and Sections under Nangod 1960**

Chiefdom	Population	Section	Population
1 Nangodi	3,364	1 Kalini	971
		2 Guosi	313
		3 Nkunzesi	105
		4 Nyaboka	102
		5 Yakoti	600
		6 Soliga	734
		7 Nakpaliga	445
		8 Gongo	94
2 Dusobilogo	170	9 Dusobilogo	170
3 Zoa	854	10 Zoa Nayiri	389
		11 Bariyambisi	465
4 Kongo	1,888	12 Adusabisi	1,134
		13 Asinabisi	754
5 Logri	864	14 Logri Nayiri	475
		15 Aringon	389
6 Dagliga	611	16 Dagliga Nayiri	234
		17 Gaagin	377
7 Damologo	716	18 Damologo Nayiri	456
		19 Zook	260
8 Pelungu	1,491	20 Pelungu Nayiri	994
		21 Vea Dabok	497
9 Tindongo	1,145	22 Tindongo Nayiri	504
		23 Ntoog	641
10 Zanlerigu	2,530	24 Zanlerigu Nayiri	957
		25 Namoog	357
		26 Zoagin	478
		27 Gani	427
		28 Asongo	311
TOTAL	13,633	TOTAL	13,633

* Populations estimated from the 1960 Ghana Census of Population, Vols. 1 and 2, and a Northern Region map of census enumeration areas, 1 : 250,000, designed by the writer.

from 47 to 11, and population fell by 67 per cent; for Nyaboka (Section 4) the figures were 62 to 28 houses and 75 per cent. Population density in the frontier sections is about 50–100 persons per square mile, as compared with as high as 1,000 in some interior sections.

When a house is abandoned, its roof timbers are usually removed, so that within a few years the torrential wet-season rains wash its mud walls to the ground. As a result, evidence on the ground is easily overlooked. Air photographs, however, show no lack of evidence of former occupation. In more recently abandoned areas, nearer the present edge of settlement, traces of the typical radial system of footpaths can be seen. There is also evidence of sheetwash erosion in overcultivated areas now abandoned. In these areas tree growth is still restricted, decades later. But the clearest evidence lies in the different rates at which woody vegetation recolonises abandoned farmland. The area immediately around each house customarily receives the greatest amount of manure; consequently, when abandoned, it supports a more vigorous growth of natural vegetation. Clumps of thicker vegetation thus reveal former house sites, particularly where they contain baobab trees, which commonly serve as domestic shrines.

Part of the abandoned area now lies in the Red Volta West Forest Reserve (Fig. 19.3). Although the order establishing the reserve was not promulgated until 1956, it was demarcated as early as 1941. At that time there were no settlements within the proposed reserve. Indeed, the boundary was deliberately drawn to exclude all habitations, and it therefore gives a good idea of the frontier of settlement in 1941. A Forestry Department 'selection report' of 1950 refers to abandoned and derelict compounds and rubbish mounds throughout the reserve.[18]

Details of settlement retreat in Sakoti, a typical locality, are presented in Figure 19.4. Sakoti, lying immediately south of Nangodi, was chosen because the availability of 1 : 10,000 air photography, taken in 1949, provided a convenient basis for fieldwork. It is known locally in Sakoti that settlement once extended to the bank of the Red Volta. Figure 19.4 reveals a total absence of settlement near the river today, and continuing abandonment of habitations. Between December, 1949, and the writer's ground survey in April, 1963, no fewer than thirty-six houses were abandoned in the small area mapped. One house was in process of evacuation during the period of survey.

It would seem from discussions with elders along the present edge of settlement in Nangodi that the bank of the Red Volta was settled as late as 1918, or possibly a little before. We may calculate, then, that around Gongo, north of the main road, settlement has retreated at a mean rate of one mile every seven years; south of the road around Nkunzesi and Nyaboka, where farmland has been relinquished more slowly, the mean rate is one mile every twelve to fourteen years. As farming declines, scrub and trees advance to recolonise the abandoned area, and game increases. The Nangodi-Naba's house in Kalini, once centrally situated, now stands at the eastern extremity of the chiefdom; this is true also of the Sakoti-Naba's house.

All along the retreating frontier of settlement there is a pervading atmosphere of decline and decay. Few of the houses are well maintained; although land is abundant, farms are not large and tend to be neglected because of limitations in the labour force, many of whom are blind. To the lay observer, nutritional standards seem to be low, and the incidence of minor infections, such as sores, is much higher than in the interior. When one visits these small, isolated, and dying communities, one cannot help feeling the contrast between the vigour and energy of the central and

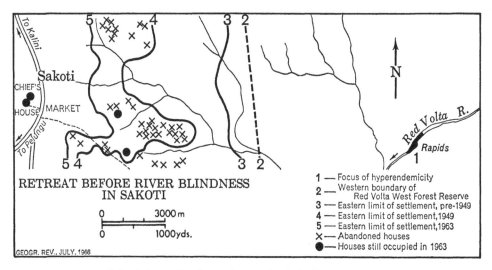

FIGURE 19.4 Sakoti, immediately to the south of the Nangodi traditional area, is examined here because of the availability of air photography of suitable date and scale to show the process of retreat from the river (it is known locally that settlement once extended to the bank of the river). In the area between lines 3 and 4, abandoned footpaths are visible on air photographs taken in December, 1949 (scale 1 : 10,000), though most of the houses had washed away. Crosses between line 4 and line 5 (limit of settlement at time of author's ground survey in April, 1963) represent 36 houses abandoned during the 13½ year period. Of the two houses shown as 'still occupied', the lower one is a resistant outlier of settlement; the upper one was in the process of being abandoned at the time of the author's survey.

western sections, such as Soliga and Zanlerigu, and the lethargy and quiet resignation on the border of the blind area.

The assumption implicit in our discussion so far that river blindness is, and always has been, the sole reason for abandonment of farmland may now be questioned. Other possible causes of depopulation are slave raiding, soil erosion, and other riverine diseases. There are old men in Nangodi today who remember the slave raids of Babatu, who pillaged north-eastern Ghana just before the turn of the century. However, there is little reason to suppose that Babatu and his forces operated selectively in riverine areas;

neither can one assume that the areas attacked were completely devastated and that no population survived. Recent historical research suggests that slave raiders' policy may be likened to the collection of honey – never in excess of the natural rate of replacement.[19]

Soil erosion is certainly a factor, but it also does not appear to be more serious near the river than elsewhere. Although erosion scars are evident in the abandoned zone, they are not more numerous there than in present areas of habitation. Alluvial land beside the Red Volta possesses a high level of fertility; there is thus no question of soil exhaustion. Those who had left the blind area all maintained that the land was in good heart.

As for diseases other than onchocerciasis, it is likely that one at least, trypanosomiasis (sleeping sickness), has operated in favour of the evacuation of riverine areas, since its tsetse-fly vector prefers a riverside habitat. The disease is not significant in Nangodi today, but the possibility of epidemics in the past cannot be ruled out. Undoubtedly outbreaks of sleeping sickness have periodically devastated riverine belts in many parts of tropical Africa,[20] and there is no reason to suppose that northern Ghana has escaped such visitations. Survivors in Nyaboka spoke of *Simulium* flies and tsetse flies, and of houses near the river suddenly dying out here and there because of 'sickness' that could be suggestive of trypanosomiasis. Other diseases that could have devastated entire areas if allowed to run unchecked are smallpox and cerebrospinal meningitis, but their geographical distribution is not confined to riverine areas.

Although the possibility of outbreaks of sleeping sickness cannot be excluded, the weight of contemporary evidence suggests that prime responsibility for the evacuation of riverine areas rests with onchocerciasis and its 'diabolical little vector',[21] *Simulium damnosum*. In 1955 a Medical Field Unit made a survey of the distribution of onchocerciasis in Nangodi, the results of which are presented in Figure 19.5. According to the evidence of skin snips, average infectivity was 60 per cent, ranging from 31 per cent in Zanlerigu to 83 per cent in Zoa (Table 19.2). Heavy infectivity, measured by the proportion of subjects with nodules, shows a maximum concentration along the frontier. In fact, the linear distributional pattern of acute onchocerciasis readily explains the decline along the present edge of settlement.

A question of primary importance is the absolute limit, if any, to which settlement will have to retreat. The whole of Nangodi lies within the flying distance of an unparasitised vector, and already all parts are infected. The problem therefore is: To what extent will *heavy* onchocercal infection continue to spread? It is known that the vector is shade-loving, and thus the recolonization of abandoned farmland by woody vegetation may be helping the spread of the fly from the river. Furthermore, the increase of game in the abandoned area provides a reservoir of the disease. Erosion rills and gullies, which are not uncommon in the settled area, create suitable oviposition sites in the rainy season; hence breeding foci are not confined to the Red Volta and its main tributary streams. In addition, the already high rate of human infection ensures that a large proportion of the vector population will become infected with *Onchocerca volvulus*. It is difficult to conceive of an absolute limit of retreat in Nangodi, unless the vector population is reduced or the parasitic cycle is broken by large-scale administration of filaricides to the human population. On the other hand,

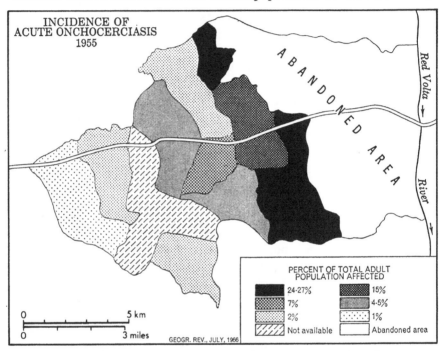

FIGURE 19.5 Estimated proportion of adult population with subcutaneous onchocercal nodules in 1955 (see Table 19.2). Note the heavy concentration of infection along the edge of the abandoned zone.

the entomological evidence revealed that breeding flies are more closely confined to rivers than the possible flight range of the vector would suggest. In Nangodi the principal breeding ground, dwarfing all others in importance, remains the Red Volta, and the problem therefore relates basically to distance from that river.

One might imagine that the evacuation of riverine areas was associated with a large displacement of population, but after visiting the frontier of settlement and questioning headmen in interior sections, the writer has concluded that only a very small proportion of the original population left the riverside areas. Most people, it appears, died where they were. Babil, the headman of Kalini, recalled by name thirty-four houses that had died out in his lifetime in the area now abandoned. In Gongo, a sorely stricken 'relic' section, the headman, Adam, named twenty-one houses known personally to him in which all the inhabitants had died. Among the interior sections, in Soliga and Tindongo Nayiri seven houses and four, respectively, had returned from the river, but in most sections (for example, Damologo, Zook, Zanlerigu) none had returned. One's overriding impression is that,

TABLE 19.2 *Incidence of Onchocerciasis among Adults in Nangodi, 1955**

| Area | Visually examined | | | Skin snips | | |
	Total	Nodules[a]	%	Total	Positive[b]	%
Kalini[c]	646	156	24	300	211	70
Yakoti	268	13	5	136	85	63
Soliga	318	21	7	151	84	56
Nakpaliga	540	80	15	266	162	61
Gongo	51	14	27	26	18	69
Dusobilogo	169	3	2	56	42	75
Zoa	448	9	2	143	118	83
Kongo	671	30	4	181	121	67
Dagliga	98	2	2	46	23	50
Pelungu	576	12	2	174	115	66
Zanlerigu	1,169	8	1	338	106	31
	4,954	348	7	1,817	1,085	60

* Based on unpublished data kindly provided by Mr. A. Abatey, Field Superintendent, Gambaga, by courtesy of Specialist i/c Medical Field Units, Kintampo, Ghana. Adults are persons 20 years of age or older.

[a] Presence of nodules (onchocercomata: subcutaneous breeding sites) indicates heavy infection; see Figure 19.5

[b] Positive skin snips indicate infection, degree not determined. A single skin snip was taken from each subject's calf and examined microscopically for microfilariae.

[c] Kalini here includes Guosi, Nkunzesi, and Nyaboka.

through sickness and disease and the inability to farm properly because of failing sight, riverside populations have largely died out rather than removed en masse.

It seems probable that until about ten years ago there was no awareness of a causal relationship between riverside farming and a higher incidence of morbidity. Recurring sickness and death in a house would lead eventually to its being condemned as 'bad,' and hence necessitate removal to a fresh site for those well enough to make the effort. Today the knowledge that flies bring blindness is widespread. Some farmers in Nakpaliga, for example, although short of land, will not make 'bush' farms because of fear of becoming blind; the view is held that a couple of seasons of farming in the bush are enough to bring blindness. Nevertheless, the bulk of those living on the frontier today, in Gongo, Nakpaliga, Kalini, Guosi, Nkunzesi, and Nyaboka, are disinclined to move, lacking either the energy or the will, or indeed the opportunity to establish farms in the already over-crowded interior. Some members of a family may move, but most seem to be stoically resigned to a fate of poverty, malnutrition, blindness, and disease, with the forces of an inhospitable environment gradually gaining the upper hand.

For a population seasonally hungry and never far from starvation, blindness, striking a man in his prime, is a calamity. A man who could normally support a family by his efforts with a hoe in the rainy season is incapacitated and becomes a burden on the community. Less food is produced, and already precarious levels of nutrition are lowered further. Spreading blindness is causing the abandonment of valuable agricultural land, while, in the interior, population pressure is already far in excess of the carrying capacity of the existing agricultural system. Two results of this pressure are soil erosion in many overworked areas and the emigration of the younger men, key members of the labour force, to southern Ghana.

A HYPOTHESIS OF CYCLICAL ADVANCE AND RETREAT OF SETTLEMENT IN RIVERINE AREAS

Local investigation in a preliterate society, where there are few if any documents, presents special problems that need not be emphasised here, except perhaps to mention the obvious importance of sifting and cross-checking the oral evidence one collects. Old people can still be traced in Nangodi who remember spending their childhood in settled areas on or near the Red Volta. From attempts to date their life histories against well-known events such as Babatu's slave raids (1890's), the advent of the Europeans (treaty with Mamprusi, 1897), and the eviction of the Tallensi from the Tong Hills (1911) it appears reasonable to assume that the bank of the Red Volta was occupied at the turn of the century and that the retreat did not begin until about 1915–18. However, possibly more interesting than the retreat is the evidence for an advancing wave of settlement into riverside lands two decades or so earlier. This advance seems to have begun about 1890 and to have culminated in 1895 at the time of a major famine. Families entered the riverside bush from all parts of Nangodi, and from districts outside such as Bongo, Bolga, Navrongo, and even Mamprusi. In the words of the headman of Gongo (translation): 'All rushed down here for food. Because of the hunger they came. People had lots of cattle and big farms. When my people came, it was a nice place. There were no flies.' Gongo was, by repute, one of the biggest markets for miles around. Now it comprises only twelve quietly decaying houses.

The present-day structure of lineages (a lineage is an agnatically related group of families forming a segment of a clan) also suggests an earlier migration toward the river. Lineages are much reduced along the frontier of settlement, ranging in size from one to fifteen houses (Fig. 19.6). Significantly, also, there is a larger number of mixed lineages along the frontier. Nyaboka, with eleven lineages, only three of which are exogamous, is a good example. In interior Nangodi lineages are fewer and larger, and exogamy is frequently 100 per cent.[22] By this token, social structure tends

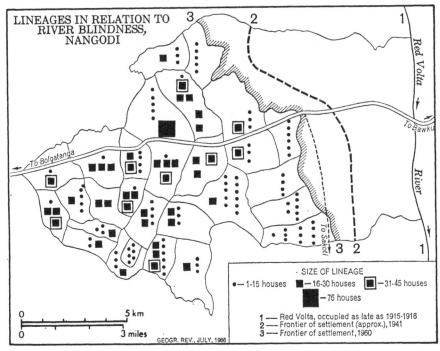

FIGURE 19.6 Number and size of lineages in relation to river blindness. Data for the approximate frontier of settlement in 1941 are from Forestry Department files, for the frontier of settlement in 1960 from air photography of that date.

to be homogeneous in interior 'core' areas of settlement and splintered and diverse toward the river. Migration from the interior, and mixing of settlers along the bank of the river, appear to be the obvious explanation. Chiefs and headmen confirm this. With the exception of Zoa, which customarily 'forbids' emigration, settlers from every clan, section, and chiefdom of Nangodi (and many outsiders besides) moved into the riverside bush to make farms. The great majority did not return. In the light of the foregoing, the notion that spreading blindness has increased population density in central, watershed areas needs to be modified. In fact, population pressure in central areas, in the face of crop failure and hunger, caused the

K

migration to the riverside in the first place; and very few of the early settlers returned to the interior.

It seems fairly certain that a cycle of settlement is nearing completion. At the end of the nineteenth century, motivated by famine, perhaps preceded by a series of poor harvests, settlers from a wide area moved into empty, riverside bush; some even settled on the banks of the river and drew their daily water from it. They prospered for a while, possibly for twenty to twenty-five years (1890–5 to 1915–18 approximately), and then disease began to take its toll, wiping out whole communities, weakening others and reducing their capacity to work and to feed themselves adequately, so that slowly, inexorably, the process of abandonment came into operation. Beginning along the Red Volta, where the incidence of disease was greater, land has been abandoned at the rate of one mile every seven to fourteen years for the past forty-five to fifty years. Nearly two-fifths of Nangodi has now been deserted, and the retreat continues. Although a combination of riverine diseases (including sleeping sickness) may be responsible for this retreat, the primary agent is, indisputably, river blindness.

If this cycle of advance and retreat of settlement in riverine areas has occurred once, as can be demonstrated, it seems possible, indeed likely, that it has occurred before, since historically famine and disease are ever present, like repellent poles in a magnetic field – hunger pushing in one direction, disease in the other. Land pressure and prolonged hunger would drive settlers into riverine areas, fertile, empty, and deceptively inviting. After a relatively short time the forces of disease gain the upper hand, and as settlements die out, advancing bush recolonises the deserted farmland, and the cycle is complete.

A critical assumption in this hypothesis is that at some stage the incidence of disease diminishes sufficiently to halt the retreat of settlement. Equilibrium could be established in several ways. The fly population could decrease for some unknown ecological reason; certainly, its numbers would decrease with increasing distance from optimum breeding sites. The human reservoir of infectivity would also decrease, simply by the death of the worst-infected subjects. When settlers advanced to the Red Volta, toward the end of the nineteenth century, conditions were apparently favourable, and we may therefore assume that a state of equilibrium existed, with the fly population and/or its infectivity in a 'quiescent' phase. On this ecological question, however, there is need for further investigation.

Another consideration is that if the cycle has operated historically over the centuries, settlers of each advancing wave could easily have been unaware of the hazards of farming near the river. Even where legends about blindness and disease had been passed down over the generations, their injunctions would carry little weight when a hungry settler visited an apparently fly-free and healthy area. Modern awareness of the risks of riverine habitation, together with the development of famine relief schemes, makes it unlikely that such cycles will operate freely again. In the light of African aspirations for major agricultural development, the case for future permanent reclamation of the river lands is strong.

REFERENCES

1 RODGER, F. C. (1959) *Blindness in West Africa*, p. 81. London.
2 BURCH, T. A. (1961) 'The ecology of onchocerciasis'. In J. M. May (ed.) *Studies in Disease Ecology*. New York.
3 RODGER, F. C. (1959) *op. cit.*, especially pp. 212–22.
4 Calculated from CRISP, G. (1956) *'Simulium' and Onchocerciasis in the Northern Territories of the Gold Coast*, p. 61, table 19. London.
5 RODGER, F. C. (1959) *op. cit.*
6 *Ibid.*, p. 225.
7 *Ibid.*, p. 234.
8 CRISP, G. (1956) *op. cit.*, *passim*. For an excellent bibliographical survey of the entomological aspects see BURCH, T. A. (1961) *op. cit.*
9 CRISP, G. (1956) *op. cit.*, p. 99.
10 DAVIES, L. (1961) 'Ecology of two prosimulium species (diptera) with reference to their ovarian cycles'. *Can. Ent.*, **93**, 1113–40.
11 CRISP, G. (1956) *op. cit.*, p. 122.
12 The investigation is being made by the Onchocerciasis Research Station, National Institute of Health and Medical Research, Bolgatanga, Ghana, under the direction of Mr. T. McRae, to whom the writer is indebted for interesting discussions of the problem. The idea has not hitherto been pursued because it was believed that the rains came from the east. Line squalls, the commonest form of rainy season precipitation, do in fact travel from east to west, but they are wavelike disturbances that develop in the T_m air and in the overlying T_c air. Rainstorms also strike from the east, but the overall pattern of precipitation shows a seasonal advance from south to north behind the advancing intertropical convergence, as one might expect.
13 CRISP, G. (1956) *op. cit.*, p. 128.
14 RIDLEY, H. (1945) 'Ocular onchocerciasis, including an investigation in the Gold Coast'. *Br. J. Opthal.*, Monograph Suppl. No. 10.
15 WADDY, B. B. (1949) 'Onchocerciasis and blindness in the northern territories of the Gold Coast'. Cyclostyled report. Ministry of Health, Accra.
16 CRISP, G. (1956) *op. cit.*; RODGER, F. C. (1959) *op. cit.*
17 The following discussion is based on part of a survey of the human geography of Nangodi, recently completed by the writer, which included clan structure, settlement patterns, population pressure, agriculture and seasonal hunger. Field investigation comprised reconnaissance sketching, detailed survey of selected areas, air-photo interpretation, and interviews with chiefs, section leaders and elders. With respect to the problem of river blindness, all settlements along Nangodi's retreating frontier were visited.

18 Forestry Department files by kind permission of the Senior Assistant Conservator of Forests, Navrongo, and the Chief Conservator of Forests, Accra.

19 Personal communication from WILKS, I. G., Professor of History, Institute of African Studies, University of Ghana.

20 MAY, J. M. (1961) 'The ecology of African trypanosomiasis'. In *Studies in Disease Ecology*. New York; DESHLER, W. (1960) 'Livestock trypanosomiasis and human settlement in north-eastern Uganda', *Geogrl. Rev.*, **50**, 541–4; MORRIS, K. R. S. (1965) 'The ecology of sleeping sickness', *Discovery, Lond.*, **26**, 23–8.

21 WADDY, B. B. (1949) *op. cit.*, p. 3.

22 Data from the writer's survey of clans, details of which are not discussed here.

20 Modelling the Geographic Epidemiology of Infectious Hepatitis

A. A. Brownlea

Attempting to build a model to simulate a range of spatial and temporal features of infectious hepatitis incidence in the Wollongong system of settlements between 1954 and 1970 means handling a highly dynamic and complex situation.[1-6]

This coastal city of 180,000 people, some fifty miles south of Sydney, capital city of New South Wales, has undergone several major transformations in the last two decades. First, an economic transformation saw the former economic base of the settlement change from dairying and coal mining to heavy industry (steel making, engineering, chemical plants) and tertiary activities (university and colleges, growth in commerce and transport). Second, demographically, the age structure of the communities has changed from relatively low to relatively high proportions of people under 21 years, and the proportion of single males has also risen. In addition, the proportion of people whose birthplace was outside Australia has risen, and in many communities exceeds the local-born. These two transformations, economic and demographic, have been the direct regional products of Australia's rapid industrial growth and immigration policy. Third, there has been a transformation of the settlement pattern. The older independent townships based on collieries, farm servicing or recreation (holiday towns, retirement villages) have largely lost their identity as suburban sprawl from Port Kembla and Wollongong has engulfed them, turning them into enclaves of old shops and grey-brown houses in a surround of newness and pastel colours, representing the homes of the 10,000 people arriving in the area each year. This suburban frontier has moved across the former dairy pastures, particularly on the flatter, more open areas south of Port Kembla and around the shores of Lake Illawarra. But, even so, a more subtle segregation is to be seen; a segregation of housing by nationality. The 20,000 British migrants are most widely spread, but there is a strong concentration of them in Berkeley; the 25,000 other Europeans have nucleated – the Italians and Greeks live close to their work in Warrawong and Warilla, the Germans and Dutch live by the railway, away from their workplaces, in Oak Flats and Fairy Meadow areas. The Yugoslavs have clustered in parts of Cringila. The young Australian families are minority groups in some of these suburbs. There is still another subtle segregation taking place: the higher socio-economic, white-collar worker groups, mainly Australian-born are building their homes on the more expensive, often waterborne-sewered foothill subdivisions (Figtree Heights, Mount Ousley), leaving the lower income, blue-collar worker families on the poorly drained, unsewered residential subdivisions. However, despite any outward appearances of independence, these diverse communities are now linearly linked together in an economic system that has, as its focus, either the steel plants at Port Kembla or the commercial core of Wollongong: the whole system feels the effect of the fortnightly paydays of the steelworks.

Amid sprawling suburbia and government housing settlements, community-closing neighbourhood and regional shopping centres have been built in Corrimal, Fairy Meadow, Warrawong, Dapto and Warilla, often growing onto older shopping areas. These centres polarise traffic flows and social interaction, but the major 'melting pot' remains, both literally and figuratively, the Port Kembla steel-producing and allied plants. The fourth transformation has been in status, from a 'country town' to a centre of both regional and national importance. The beaches of Wollongong and Kiama are the tourist attractions and holiday resorts of many Sydney-dwellers, but their importance has been eclipsed by the growing national significance of the steel plant (5·4 million ingot tons annually by the mid-1970's) and the growing regional significance of Port Kembla harbour ($40 million developments under way). This new status has changed the volume and direction of commercial and population flows both in New South Wales at large and in the Wollongong network itself. A major express highway proposal aims at linking Newcastle, Sydney and Wollongong to facilitate commercial interchanges. Thus, the old era of stability and relative isolation has gone, probably forever: in its place has come rapid population growth tied to migration and industrial development, and an emerging national significance. In sum, there is a continually changing ecological diversity in the area: a distinctive epidemiological environment.

These transformations have produced problems. Among the more significant ones for the present study are those related to environmental pollution, overcrowded schools, badly sited housing estates and the widespread absence of waterborne sewage systems. Lake Illawarra, used for swimming, fishing, prawning and boating, receives heavy run-off from creeks, gutters and drainage lines that carry faecally contaminated septic wastes from an increasing number of unsewered homes on the heavy clay soils around its shores. The concentration of faecal coli rises to dangerous levels after rain and at times when sand drift closes the entrance (and exit) of the Lake. Industrial pollutants add further hazards to the recreational use of the Lake. Schools cannot easily absorb the growth in numbers of children aged 5–8 years: in 1966 the Mount Warrigal area was dairy pasture, in 1967 a school was opened there and enrolled 380 children from the newly-built housing estate. This school's population reached 751 in 1968, 833 in 1969 and exceeded 1,000 in 1970! Serious pressures are placed on septic systems under these crowded conditions. The general monotony of the house style, the rapid build-up of pseudo-communities, the high proportion of very young children, the absence of shops and community facilities, the newness of many to the Australian scene, and the frustrations associated with all new housing estates create a distinctive ecological situation with strong implications for community health. Industry continues to expand, and with it, the demand for labour. The coastal lagoon that gave refuge from heavy seas to the early discoverers is now filled with industrial waste and provides the site for more factory construction. Each new factory job creates several other positions in the city, so that houses continue to spread across the heavy clay soils that once discouraged the early farmers, but now threaten the health of families that live on them through their inability to allow septic tank wastes to clear by percolation from the allotments.

These are 'green pastures' for nurturing infectious hepatitis: the viral causative agent is transmitted from person to person in faecal particles, and is able to maintain itself for long periods under a range of environmental conditions. But, much still needs to be known about the causative agent: it has not been isolated and cultured, and there is no immunising vaccine available. Further clues to its nature might lie in its epidemiology, but, again, much of the evidence is contradictory.

Koff and Isselbacher[7] have reviewed the changing concepts in the epidemiology of infectious hepatitis. Krugman[8] and Reikowski[9] both stress the significant role of the anicteric (i.e. no jaundice symptom) case in spreading the disease, and Constantinesco[10] questions the accepted permanent immunity conferred by the disease. Analysing diffusion under these conditions becomes highly complex. Moreover, a range of 'vectors', including oysters, pets, food, milk and water supply, has been incriminated.[11, 12, 13, 14] These 'vectors' are all relevant in the Wollongong situation with its strong emphasis on recreational facilities. The contribution of the school to the spread of infectious hepatitis is unclear: Knight[15] blames it, but Mosley[16] disagrees. Instead, a variety of environmental factors has been proposed: the spring thaw, heavy rainfall, heavy soils, severe drought and so on. These suggest that the virus is a very hardy one. There is no complete agreement on the relationship between socioeconomic status and the incidence of infectious hepatitis: separate studies show that in California case rates have been highest in higher socioeconomic groups, whereas in Rochester the reverse held true and in Baltimore[17] there was no relationship at all between income and incidence. However, the rising current world incidence may be a product of contemporary sociological changes,[18, 19] and several studies[20, 21, 22] relate incidence and population density. Other studies stress the importance of the age structure of the community.[11, 16, 23] There has been a dearth of long-term studies of the epidemiology of the disease in a single community; one study[24] only covered a two-year period.

THE WOLLONGONG DATA

The Wollongong study was deliberately made long term (1954–69) to analyse the diffusion process, the operation of ecological parameters, the contribution of population change and to minimise the effects of chance elements, as far as possible, leading subsequently to model formulations of the disease behaviour. The basic data were the records of notified cases of infectious hepatitis. Since 1954, all medical practitioners have been required to report all cases of the disease to the Health Department, with details of the patient's age, sex, occupation, family, contacts, and so on. Approximately 2,000 cases were reported in the study area between 1954 and 1969. Now, these data have limitations: not all cases are notified and, at times, the diagnosis is incorrect. Furthermore, it is difficult to determine the representativeness of the actual notifications. Some doctors notify all cases; others only some of their cases. Since all data have limitations, field-work, in this instance, was able to supply a large number of the 'missing' cases in selected areas. Moreover, certain tests of the data were encouraging. For example, the normal expectation is equal incidence in the sexes, and for Wollongong, over the period, 333 cases were reported, of whom 167 were males and 166 were females. Nevertheless, it is probable that the higher socio-economic group cases are under-represented in the data, and, after intensive mass media publicity, it is likely that a higher reporting percentage occurred in the years 1963–6. As a working rule it was taken that a single notification represented at least the *presence* of the disease in the community and that it also provided some guide to the *sequence* of outbreaks in the area.

For the purposes of this review, six features of the notification pattern can be mentioned. First, all communities notified cases, but at different rates per thousand population. A cumulative summary is given in Fig. 20.1A, showing the high rates occurring in Unanderra, Shellharbour, Dapto, Coledale and Kiama. Second, there were cyclic fluctuations in annual totals, with peaks in 1956, 1961 and 1966, with case incidence rates approximating 2·0 per 1,000 population per annum, falling to a relatively uniform rate of 0·3 per thousand population during 'lean' years. The highest annual total was 311 cases in 1966, but the case incidence rate was highest in 1956. Third, the monthly incidences have been highly variable, but the modal months of highest notification in any year are September

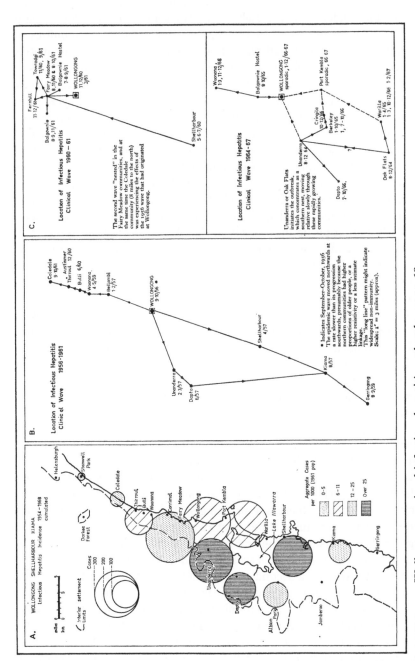

FIGURE 20.1 Wollongong regional infectious hepatitis outbreaks, 1954–68.

and October. While the major settlements can have large outbreaks in any month, the tendency in the Lake Illawarra communities is for a slight winter maximum of cases. In the unsewered beachside communities of Austinmer and Thirroul outbreaks most regularly occurred approximately one month after the school vacation periods with their strong influxes of holiday makers and visitors. Fourth, almost 70 per cent of the cases were aged under 21 years, but the largest single group was the 5–9 year olds. These children, who constituted only 20 per cent of the area's population, provided 52 per cent of the notified cases. By Salk inoculation or Sabin ingestion they had been spared poliomyelitis in the outbreaks of 1960–1 but they had no protection against infectious hepatitis: polio has been replaced by infectious hepatitis in a kind of *sequent morbidity* situation. Fifth, there was an equal incidence in the sexes: 897 males, 887 females in the 1954–66 period. Finally, about 15 per cent of the annual cases came regularly as multiple notifications in the family, e.g. in one year six members of the same family contracted the disease. Not surprisingly, a correlation of $+0·97$ (significant at the 0·1 per cent level on Student's t test) was found between the size of the outbreak (number of notified cases) and the number of cases occurring as multiples in families.

The regularities of the notification record were matched in several ways by an orderliness in the patterns of spatial incidence of the disease. First, infection tended to proceed somewhat regularly along a street, with the earliest cases located around the prime source of infection. An overflowing septic tank at the local school allowed effluent to seep into a creek used as a swimming hole in the hot summer months by children who later became patients. Social and other school contacts diffused the infection further along the street (see Fig. 20.2B). In another residential subdivision in Warilla as the streets lengthened with house construction, the new arrivals became the fresh victims, particularly in low-lying, badly drained, un-sewered parts of the subdivision. Second, infection is apparently highly selective in terms of people, and because particular people are not randomly located residentially, there is a matching spatial selectivity as well. Within the Corrimal community, on at least two separate occasions, there was a regular diffusion pattern: the first cases clustered in the modest homes on the low-lying unsewered sites around the railway station (the link to Port Kembla steelworks), a second nest developed later around the shopping centre on the highway, a third outbreak took place around the school and a final group of cases, fewer in number now, was located in sewered homes of the highest socio-economic class, located attractively on the well-drained, sandstone foothills overlooking the rest of the community. Third, there are interesting spatial parallels in notifications in the Berkeley and Kiama communities: what might be called a 'yo-yo' or 'swash-backwash' pattern. In Berkeley the initial outbreak occurred under crowded conditions in one block in the English-migrant hostel, the second took place in the surrounding Berkeley community, and the final outbreak was again located in the hostel, but now in another block. Similarly, in Kiama, the first outbreak was sited in the township itself, followed by a subsequent diffusion of cases to the Gerringong farming community and a final return of cases in the Kiama township. This core–periphery–core sequence has occurred in other areas as well, e.g. Dapto–farmlands–Dapto. Fourth, the Berkeley situation can be set into a wider spatial context: the regular succession of outbreak foci through the shore settlements of Lake Illawarra. For reasons not clear, but maybe related to the living and density conditions that obtain in migrant hostels, or to differences in immunity between different nationality groups (e.g. Italy has had a recent history of infectious hepatitis epidemics, so that Italian migrants might bring relatively higher levels of immunity when settling in these parts), or to virus-infected windblown spray from faecally polluted Lake Illawarra moving northwards before winter southerlies and so on, the earliest outbreaks occur on the northern side of the Lake, and closest to the steelworks complex, then occur progressively around the western and eastern shores to the southern communities. In other words, there is some evidence of a regular spatial sequence in outbreak timing. Finally, in this progression there is a tendency for relatively fewer cases to occur on the sandier soils close to the coastal dunes, where ground drainage of domestic wastes is more efficient, than on the more easily waterlogged heavy clay soils away from the coastal edge.

Furthermore, there is a distinctive 'ecology' of outbreak behaviour and spatial patterns in years of light and heavy incidence. The ecology of *light* incidence years is characterised by a combination of, naturally, a small number of notifications at a low incidence rate, a concentration of cases in adults rather than children, sometimes a skewed sex incidence, cases occurring without any marked seasonal rhythm, and an apparently irregular spatial distribution, as if a process of 'spatial infilling' were taking place. The pattern in *heavy* incidence years, on the other hand, is almost the converse of the light incidence years: a high case incidence rate, an occurrence of most cases in children, an equal sex incidence, a discernible seasonal behaviour, and a definite spatial concentration in one (or more) outbreak focus that develops relatively quickly during the progress of the epidemic. This marked spatial localisation of cases in heavy outbreak years, then, complements the scattered pattern in minor outbreak years.

MODELLING THE EPIDEMICS

Now, these local regularities are evidence of a much broader pattern of orderly disease diffusion through the settlement network. It is assumed that the actual diffusion process is observable as a 'clinical front' of overt cases, analogous to meteorological observation of the passage of a cold front. It is assumed, further, that the arrival of this diffusing or moving clinical front is coincident with a marked rise in the number of notifications of infectious hepatitis in the community. The location, over time, of this moving clinical front can then be plotted from the notification record. These premises form the essential theoretical or conceptual basis for generating the network diffusion models.

Before the actual network models could be constructed, some assessment had to be made of random elements in the notification record. Poisson tests of randomness were applied to the annual notifications from each suburb, most confidently, of course, where there had been little change in the community population over the study period. The Stanwell Park community, a small holiday resort on the northern extreme of the settlement network, may serve as an illustration. Now, nine cases of infectious hepatitis were reported from Stanwell Park between 1954 and 1966, giving an actual annual mean of 0·7 cases. Assume that the distribution of actual cases is Poisson in form. The actual annual mean (0·7) becomes the mean and variance (by definition) of the assumed Poisson distribution. Its standard deviation would be 0·84 (square root of variance) and the expected number of cases in any year would lie in the range zero to three cases (mean ±3 s.d.). These Poisson expectations match the actual notification distribution for Stanwell Park. Furthermore, in this Poisson distribution there would be a probability of 0·5 that zero cases would occur in any year, giving an expectation of six to seven years with no notified cases: this fits the Stanwell Park record exactly. Therefore, the assumption that the annual incidence in this community is random appears reasonable. This technique was applied, with modifications, to all the suburban data to identify those individual annual notification items that fell *outside* the random range. The months of these notifications were extracted from the main body of data and plotted in a network format to show the changing origins and locations of the moving clinical front. The result is shown as Figs. 20.1B and 20.1C. It would appear that an epidemic wave originated in Wollongong in September–October 1956 (9–10/56), then moved northwards and southwards, to reach Coledale in September 1961 and Gerringong in August 1959. In the meantime, a second 'wave' nested in the Fairy Meadow communities in 1960–1 and a third wave originated in Unanderra or Oak Flats, subsequently moving through the lakeshore communities.

In addition, the rate of movement of this changing clinical front can be calculated, and expressed as miles covered per incubation period. The time unit chosen was a function of the processes involved in the disease system. There are differences in the speed at which the front appears to have moved, related possibly to density of susceptibles, linearity of the settle-

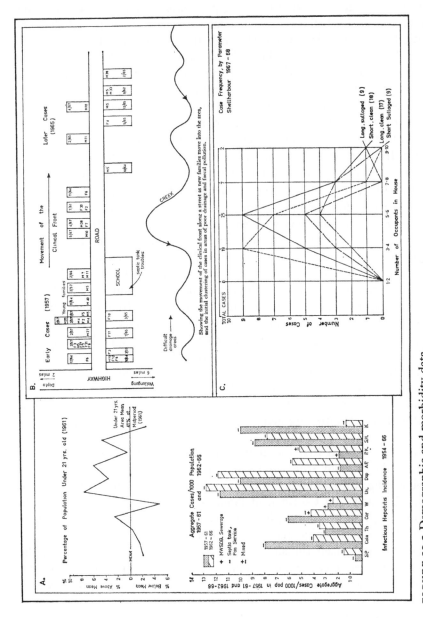

FIGURE 20.2 Demographic and morbidity data.

ment network, local environmental conditions and so on. Some ideas of the likely correlates of the diffusion rate might be gleaned from Figure 20.2A: heavier case rates appeared in young populations, in unsewered areas and in newly established residential areas. However, a more detailed and critical assessment is needed, although the possible relationships accord with these early inferences. The critical analytical procedure utilised is the *completely random walk simulation* of the clinical front. The argument is simple: deviations of the actual front from any expected location randomly determined would indicate the operation *there* of non-random constraints, or, as they are called in this paper, ecological parameters.

For further refinement then, let it be assumed that the clinical front may move, from any given origin, with equiprobability in any direction. The average distance from the origin is given by the following formulation: n times the $\sqrt{\text{'step length'}}$, where n is the number of 'steps' allowed to the moving front. For example, if the front is permitted nine 'steps', each of one mile, in any direction with equiprobability, then the front would be located on the circumference of a circle of radius $\sqrt{9}$, or three miles, described around the origin as its centre. Further, if each step can be tied to a time interval, then time elapsed, and hence speed, can be calculated. For example, if the steps were 'one mile per incubation period', then, in the previous example, the front would be three miles from the origin after nine incubation periods. This completely random walk simulation, using a diffusion step of one mile per incubation period, was developed and applied to the northern communities, with Wollongong as the outbreak origin (as it was in 1956), and expected locations compared with actual locations for the period 1956–61. The northern communities were chosen because of their contiguity and continuity in residential build-up. The southern communities are more spatially discontinuous. The comparisons are shown in Table 20.1.

It will be noted that the actual clinical front moved faster than the completely random front in those communities immediately north of Wollongong: this faster-than-random rate may be due to the presence in Fairy Meadow and Corrimal of families with young children, living in unsewered, rapidly growing housing estates, receiving considerable numbers of European migrants who are initially housed in the Balgownie migrant hostel, itself the site of serious outbreaks. Are these, then, the ecological parameters or constraints: age structure, rate of population growth, recency of arrival, nationality status, sewerage facilities? The preliminary support that had already been obtained, in Fig. 20.2A, is now strengthened. The importance, too, of the provision of waterborne sewerage systems is also clearer, because the case incidence rate fell in Corrimal after this service was provided.

The actual and expected locations of the moving fronts coincided in the outer suburbs. Perhaps there is a distance decay function applying to the diffusion rate, but, in comparison with inner suburbs, these outer communities are relatively stable in total population and generally have slightly higher proportions in the adult age groups, so that there could be some 'epidemic friction' at work in these areas. Their relative isolation is, in some ways, a protection from heavy outbreaks if distance decay processes operate. Perhaps, too, relative isolation explains the higher incidence of random cases in suburbs north of Thirroul, particularly Stanwell Park.

TABLE 20.1 *Completely Random Walk Simulation of Infectious Hepatitis Diffusion in the Wollongong and Northern Communities 1956–61*

Time Lapse in Months	Expected Distance from Origin*	Expected Suburb	Actual Suburb
0	0	Wollongong	Wollongong
6	2·5 miles	Fairy Meadow	⎧Fairy Meadow** ⎨Corrimal ⎩Bellambi
12	3·5 miles	Corrimal	
18	4·2 miles	Bellambi	
24	4·9 miles ⎫	Woonona	Woonona***
30	5·5 miles ⎭		
36	6·0 miles ⎫	Bulli	Bulli
42	6·5 miles ⎭		
48	6·9 miles ⎫	Thirroul	Thirroul
54	7·3 miles ⎭		
60	7·7 miles	Austinmer	Austinmer

 * Average distance from origin = length of step X sq. root of number of steps (assumed one step of one mile per incubation period).

 ** Faster rate may be due to local conditions, e.g. young families.

*** Both 'fronts' in step: communities with older age structure.

For the purpose of model-building, many of the basic elements of the disease system have now been identified: some random incidence, but a strong regularity in spatial and temporal patterns, conditioned or constrained by ecological parameters which include community age structure, rate of residential growth, distance from infection nodes, levels of immunity, extent of faecal pollution of the environment, linkages in the settlement network; and these parameters operate in specific sites with individual, but not always random, intensity. In this way, particular ecological situations have developed and the disease diffusion behaviour reflects the spatial arrangement of these situations: a heterogeneous 'social surface' (of locals and migrants, aged and young, mobile and immobile, home unit-dwellers and home owners, artisans and professionals, factory workers and farmers, and so on) is superimposed upon a heterogeneous 'physical surface' (of spurs and plains, sandy soils and clay soils, of lakes and lagoons, of places well drained and waterlogged, of beach and river, of forest and grassland pasture, of 'clean' and polluted, and so on).

The steps in the model-building were clear. The structure of the disease system had to be determined, and its behavioural sequences identified, in both limiting and general conditions. The spatial manifestations of this behaviour could then be incorporated, commencing simply with a single node radiating the disease on an assumed closed, uniform physical and social surface, and extending then to a single node on differentiated surfaces. Finally, the model could be further generalised to several radiating nodes in a relatively open system, and its predictive ability tested.

The most useful procedural tool in this situation is the methodology of general systems analysis. In fact, the general properties of systems[25] can be identified quite readily in the current context, and it is against this background that the models will be designed. Briefly, the disease system maintains itself energetically through the constant movement of susceptibles, carriers and a virus population. Form adjustments take place as a series of outbreaks or minor epidemics of the disease and the generation of immunes. An excess of the virus or of young susceptibles will result in an outbreak, whereas an excess of old people or the absence of the virus will result in an absence of cases. There is self-regulation in the system because the virus may maintain itself in a viable form in polluted drainage lines or in carriers until a threshold population of susceptibles accumulates and another outbreak takes place. If there is a large number of susceptibles, then the virus is able to multiply its numbers accordingly. Furthermore, the system maintains a kind of form over time: spatially, as a series of propagative waves through susceptible communities in a moderately predictable way; temporally, in a regular series of cyclic outbreaks or crises; and area differentially, by concentrating in those sectors of communities containing the largest proportions of susceptibles, or in areas of heavy soils with drainage impediments. Finally, the system appears to behave equifinally, in that once the virus is introduced into a susceptible community, almost regardless of when or how or of the proportion of susceptibles, the end product will be a protectively immune community. This condition lasts until the number of susceptibles builds up again in the normal process of population growth.

Each community has been generalised in terms of boundary conditions and flows across the boundary. An example is given in Fig. 20.3A. Within this boundary the disease system, whose structure is represented in set theoretic form in Fig. 20.3B, operates. This *first model* shows an 'energy' input of susceptibles and virus, their subsequent 'mixing' within the

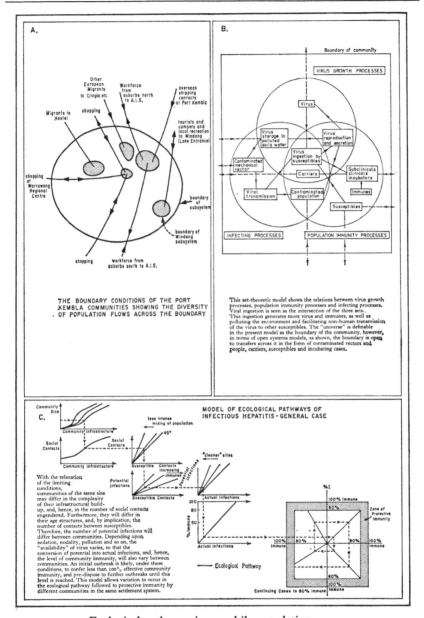

FIGURE 20.3 Ecological pathways in a mobile population.

infrastructural and mechanical vector framework of the community, resulting in the generation of infecteds, and immunes, and the continuing
'mixing' of virus and those susceptibles who had not come into contact
earlier with the virus at an infecting level, until eventually a steady state

is achieved and outbreaks cease. The model also shows 'energy' being transferred across the boundary of the community system in the forms of travelling incubating cases or susceptibles or virus outflow in polluted drainage lines or carried by mechanical vectors, such that the energy loss of one community becomes part of the energy input of another. Each community can be seen to represent a sub-system of a much larger system of settlements or communities, each with its own rate of progression to steady state, influencing the steady state condition of each other sub-system and, in turn, the whole system by these 'energy' transfers across their boundaries. Thus, some communities will be protectively immune, others at a low level of immunity and others currently experiencing outbreaks conferring higher levels of immunity in progress towards a protective level.

The *second model*, in a graphical, process-response form, shows what might be called 'the ecological pathway' of a single community moving towards its steady state condition of a protectively high level of community immunity.

The *limiting* model assumes that, regardless of the size of the community and the complexity of its infrastructure, and in a once-over change because all contacts are both potentially and actually infective, the introduction of the virus results in one massive outbreak in a totally susceptible population, and confers a completely protective level of community immunity on the population. A condition that could be regarded as one of maximum entropy would then exist, in the sense that there is a complete and continuing 'equilibrium' between the virus population and the human population, and provided that the community undergoes no further changes this condition will remain unaltered. Using the same processes, but relaxing the limiting conditions, a *general model* (Fig. 20.3c) allows for observable differences between communities and for changes in a given community in such things as the complexity of the infrastructure (influenced by stage of development, nodality, etc.), susceptibility to infection (influenced by age structure, history of epidemics, immunity status, etc.), availability of infective materials (influenced by physical properties of soil, drainage efficiency, sewage disposal system, water body pollution, community attitude to hygiene, etc.) and the degree of openness of the community which influences the number, nature and intensity of social contacts between the community and outside population. These factors acting together condition the size of the initial outbreak and the size and number of subsequent outbreaks until the establishment of complete community immunity. The ecological pathway to this point is determined by starting with population size, and deciding at each 'graph' the appropriate alternative that represents the community under review, and accepting its consequences for the next step on the ecological path.

Now, while both parts of this second model are essentially speculative and conceptual, they do indicate some of the likely lines for quantification, e.g. age structure as a measure of susceptibility, etc., and contain the seeds of predictability. The models suggest that an increase in the complexity of the infrastructure, without any population change, will increase the contacts between susceptibles, hence the size of the initial outbreak and its conferred level of community immunity, thus approximating the 'closed system, limiting model' and its *once-over outbreak*. While these ideas need to be tested rigorously, in some ways this sequence has already been observed at Warrawong: a new regional shopping centre, the community closing around it, an initial major outbreak, which establishes a high level of community immunity, and subsequent outbreaks involving a relatively small number of cases. Similar situations are likely to exist in the future at Warilla and Figtree, which are becoming regional centres.

The *third* model incorporates disease diffusion, in Hagerstrand[26, 27] tradition, or the epidemic on the ground. Assume a closed, static settlement of 100,000 people distributed uniformly by characteristic across a uniform physical surface and constituting 100 'grid' communities of 1,000 people in each, around a centrally located node which acts as the point source of the disease diffusion (see Fig. 20.4A). The *number* of infections that will occur is derived from generalised Wollongong community data and Reikowski's[9] studies of European outbreaks. Assuming that all people under 21 years of age are initially susceptible to the disease and that these comprise 35 per cent (or 350) of the population in each grid community of 1,000 people, then 'equilibrium' between the virus and these susceptibles will be achieved eventually with the following distribution of cases of infectious hepatitis: 10 per cent (or 35) overt, 35 per cent (or 122·5) anicteric, 35 per cent (or 122·5) abortive and the remaining 20 per cent (or 70) people uninfected by the virus. (In the real world situation, of course, the actual distribution will be different because of local conditions and past epidemic history. Perhaps few of the over 21 year olds may have had the disease as children and, hence, represent a big pool of potential infections in addition to the current under 21 year olds.) The *rhythm* of the outbreaks is also taken from the generalised Wollongong community records: half of the community's cases tend to come in one year, with lessening numbers over a further four-year period to reach stability with protectively high levels of immunity to infectious hepatitis. This rhythm is the basis of the 'epidemic clock' which underlies the peaks and troughs phenomena, both spatial and temporal. Furthermore, a distance decay function is built into this rhythm so that the thirty-five overt cases required for the equilibrium described above come in the following sequence for each grid community: seventeen cases in one year, then nine cases, three cases, four cases and two cases in subsequent years, or thirty-five cases over five years, but with a non-uniform decay rate because some of the incidence of the disease appears random.

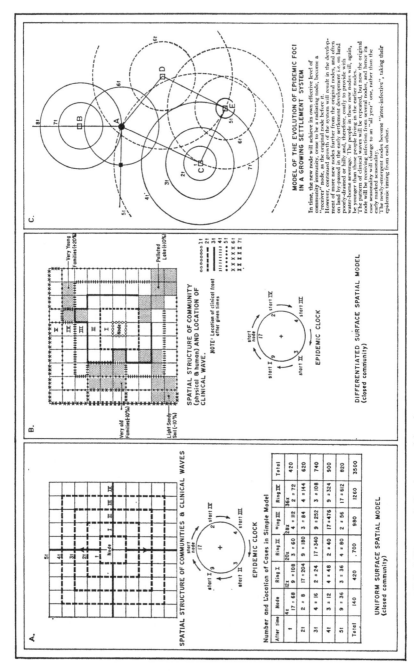

FIGURE 20.4 Modelling an infectious epidemic.

This model is shown (Fig. 20.4A) with locations of the clinical front after various time periods (t to $5t$) and the tabulation of the generated cases according to the frontal movement and the operation of the epidemic clock, e.g. after time t, the four inner grid communities around the node have had their peak notifications, yielding sixty-eight cases, and the outer rings of communities have had cases according to the distance decay function, so that the twelve grid communities in Ring 1 have each recorded nine cases (a total of 108 cases), and the twenty communities in Ring 3 have each recorded three cases (a total of sixty cases) and so on. Each community, thus, successively notifies seventeen, nine, three, four and two cases depending upon its starting point, determined by the arrival of the clinical front.

The model agrees with the Wollongong data in displaying a series of spatially changing outbreak foci (clinical front movements) and changing 'annual' notifications (after t, $2t$, and so on) which rise to a peak then fall. It further suggests that if 3,500 cases from 100,000 people were to represent the real epidemiological situation, then Wollongong's 2,000 cases out of 150,000 people would indicate that only about one case of infectious hepatitis in three occurring is actually notified to the Health Department.

Variations can be introduced into this model to align it more closely with reality. The physical surface can be differentiated and site loadings can be applied to the normal values of the epidemic clock, e.g. lighter soils in some areas with a consequent 10 per cent reduction of expected cases in those grid communities due to better virus removal, or a lake can be faecally polluted, resulting in a 10 per cent increase in expected cases in each lake-side grid community; the human surface can be more strongly differentiated, e.g. a concentration of young families can yield 20 per cent higher than normal notifications and the movement of the clinical front can also be accelerated, say, by a time period, due to increased density of susceptibles in these localities. This revised model, with the variations shaded (Fig. 20.4B), more closely resembles the Wollongong situation. It retains the general features of the earlier model but refines the movement of the clinical front by allowing it to change its rate of diffusion and to buckle as community age structure and physical surface features change with distance from the nodal origin. The original epidemic clock is retained, the loadings due to local features have been incorporated to yield the 'notifications' calculated in Table 20.2. The total number of infections in both models is similar in magnitude but distributed differently both in time and space. Some idea of the *order* of the loadings attributable to specific site features might be inferred, using the earlier model as the norm. Moreover, another spatial element has been built into the revised model: the settlement focus is now eccentrically located, and this appears to have lengthened the overall time taken for the settlement to achieve immunity

TABLE 20.2 *Number and Location of Infectious Hepatitis Cases – Heterogeneous Surface Model.*

After time	Node		Ring I		Ring II		Ring III		Ring IV		Ring V		Ring VI		Total
	Normal Cell Value* N	Total Expected Cases 1N	Normal Cell Value N	Total Expected Cases** 1N+10%	Normal Cell Value N	Total Expected Cases 10N 2N+10% 1N+20% 1N-10%	Normal Cell Value N	Total Expected Cases 14N 1N+20% 2N+10% 3N-10%	Normal Cell Value N	Total Expected Cases 10N 2N+20% 3N+10% 5N-10%	Normal Cell Value N	Total Expected Cases 14N 4N-10%	Normal Cell Value N	Total Expected Cases 3N	
t	17	17	9	73	3	43	4	80	2	58	Nil	Nil	Nil	Nil	271
2t	2	2	17	138	9	129	3	60	4	117	2	35	Nil	Nil	481
3t	4	4	2	16	17	243	9	181	3	87	4	70	2	6	607
4t	3	3	4	32	2	29	17	342	9	263	3	53	4	12	734
5t	9	9	3	24	4	57	2	40	17	496	9	158	3	9	793
6t	Nil	Nil	Nil	Nil	Nil	Nil	Nil	Nil	Nil	Nil	17	299	9	27	326
7t	Nil	Nil	Nil	Nil	Nil	Nil	Nil	Nil	Nil	Nil	Nil	Nil	17	51	51
		35		283		501		703		1,021		615		105	3,263

* N, or the normal cell value, is the epidemic clock value for that cell. It incorporates (i) distance decay along the t row and (ii) the rhythm to immunity, followed by nil cases after immunity, down the node column and (iii) the location of the clinical front diagonally across the cells.

** In Ring I, there are seven grid communities each with nine cases (= sixty-three) and one community adjacent to the polluted lake with N cases plus 10 per cent loading or 9·9 (= ten) cases, making a total of seventy-three cases in that cell.

stability. However, the actual rate at which the clinical front moves in specific sites and the case incidence loading factors still need further refinement.

A *final* formulation relaxes the closed boundary assumption to approximate more closely the Wollongong situation of rapid population growth, several changing radiating foci, different distances between the communities, and each with its distinct immunity status. This is a highly complex situation, and the model indicates more the lines of attack than the final solution. For the purposes of this paper only a late stage in the development sequence is shown (Fig. 20.4C), but the elements of the full model can be identified in it. Briefly, there are several communities of various sizes in the settlement network, but, in the early stage, the initial serious outbreak occurs at the major settlement node A, and infection radiates, as a clinical wave, in all directions (with equal probability at this stage) towards the outlying communities B, C, D and E. The effect is felt most seriously in the nearest, B, and with decreasing intensity in C, D, and E (the distance decay effect), generating in these relatively small numbers of cases, but nonetheless 'polluting' the settlement environment with the virus. Later, at A, a second outbreak occurs, but smaller now because A is moderately immune as a result of its first outbreak (the epidemic clock phenomenon). Nevertheless, this fresh wave triggers off a minor outbreak in B and a serious outbreak in the nearby community C, which has been steadily absorbing new residents because of its convenient location near the major settlement. The newly emerging node C now maintains the disease diffusion, taking over the radiating role that had belonged in the early stages to settlement A. Its clinical waves generate cases of infectious hepatitis in neighbouring communities D and E, and re-activate cases at A (the yo-yo effect), thereby assuring its eclipse as a future significant origin of disease diffusion (protective immunity level achieved), and subsequently some further cases at B. A summary of this late stage is shown as Fig. 20.4C. This process of succession is repeated as each new neighbourhood, generally of young families, develops, becomes infected, plays the radiating role till protectively immune, then 'hands over' to other emerging centres. However, one, say E (or more), of these is likely to be located on a poor site (bad drainage, etc.), and will be environmentally predisposed to becoming an endemic radiating centre. There will be constant, if weak, radiating clinical waves moving from E through the settlement network, keeping infection present, albeit locally sporadic, at all times of the year (changing seasonality phenomenon). This model approximates closely the Wollongong network diffusion (Fig. 20.1B) with changing outbreak sources, 'swash and backwash' infection waves, and has one highly interesting implication: seasonality could be a function, *inter alia*, of the spatial dimensions of the settlement network. The newly emerging nodes become inter-infective, taking their outbreak rhythm from each other, as is clearly shown in Fig. 20.4C, where the new origin has its outbreak at t, and a second surge of cases at $7t$ due to a return wave of infective material from the nearby settlement. The distance decay effect on the rate of clinical front movement remains a problem to be solved. In this way, approaches of the type envisaged by Bartlett[28] and Bailey[29] can be designed with spatial components.

The resulting model was tested against the 1968–9 notification record. The broad expectation was that infections would occur in all communities but with a marked concentration in the rapidly growing communities south of Wollongong. In the main, these expectations were fulfilled. Subsequent analysis of several specific areal non-conformities in the Shellharbour communities showed the need for more detailed knowledge of family movements, particularly to places outside the Wollongong network, and a deeper appreciation of the contribution to family susceptibility made by the stage in the life cycle: up to a point the presence of young children explains the broad pattern of incidence, but the fuller key to the micro-pattern requires knowledge of length of residence (long, short), site cleanliness (clean, sullaged), and the ages of all children and young adults in the particular families of the area, because the probability of infection appears to rise as the family members become more widely roving with increasing age (see Fig. 20.2c).

Finally, the study also suggests this lesson for public health planning: in so far as it is possible there should be a mixture of families at all stages of the life cycle in rapidly growing communities, thereby avoiding some (but not all) of the epidemiological hazards associated with high concentrations of young susceptibles, particularly in residential sites inefficiently drained and lacking a waterborne sewerage system. There has been some acknowledgement of these values in a currently planned housing development near the township of Dapto.

REFERENCES

1 BROWNLEA, A. A. (1962) 'Factors influencing the distribution of infectious hepatitis in selected areas in New South Wales'. *ANZAAS*. Sydney.

2 BROWNLEA, A. A. (1965) 'The ecology of infectious diseases in Wollongong–Shellharbour'. *ANZAAS*. Hobart.

3 BROWNLEA, A. A. (1967) 'An urban ecology of infectious disease. *Aust. Geogr.*, **10**, 169.

4 BROWNLEA, A. A. (1968) 'Towards a model simulation of infectious hepatitis incidence in a system of settlements'. *International Geographical Union Congress*. New Delhi.

5 BROWNLEA, A. A. (1969) 'The geographic epidemiology of infectious

hepatitis in an industrial city'. Conference: Australian Society for Epidemiological Research and Community Health. Adelaide.

6 BROWNLEA, A. A. (1969) 'The medical geography of infectious hepatitis'. Unpublished Ph.D. thesis.

7 KOFF, R. S. and ISSELBACHER, K. J. (1968) 'Changing concepts in the epidemiology of viral hepatitis'. *New England J. Med.*, **278**, 1371–80.

8 KRUGMAN, S. *et al.* (1962) 'The natural history of infectious hepatitis'. *Am. J. Med.*, **32**, 717–28.

9 REIKOWSKI, VON H. (1965) 'On the problems of recognising viral hepatitis as an occupational disease'. *Deutsch. med. Wschr.*, **90**, 2099–104.

10 CONSTANTINESCO, N. *et al.* (1967) 'Aspects of immunity in viral hepatitis'. *La Presse Medicale*, **75**, 1879–82.

11 LIAO, S. J. *et al.* (1954) 'Epidemiology of infectious hepatitis in an urban population group'. *Yale J. Biol. Med.*, **26**, 512–26.

12 DOUGHERTY, W. J. and ALTMAN, R. (1962) 'Viral hepatitis in New Jersey 1960–1961'. *Am. J. Med.*, **32**, 704–16.

13 HAMMETT, J. B. and HUNTLEY, R. R. (1965) 'Infectious hepatitis: report of a community epidemic'. *Southern med. J.*, **58**, 1471–4.

14 TUFTS, N. R. (1967) 'Differentiation of sources in a hepatitis outbreak'. *Publ. Hlth Rep.*, **82**, 1–8.

15 KNIGHT, V. *et al.* (1954) 'Characteristics of spread of infectious hepatitis in schools and households in an epidemic in a rural area'. *Am. J. Hyg.*, **59**, 1–16.

16 MOSLEY, W. H. *et al.* (1963) 'Epidemiologic studies of a large urban outbreak of infectious hepatitis'. *Am. J. publ. Helth*, **53**, 1603–17.

17 LILIENFELD, A. M. *et al.* (1953) 'Observations on an outbreak of infectious hepatitis in Baltimore during 1951'. *Am. J. pub. Hlth*, **43**, 1085–96.

18 STOKES, J. (1953) 'Epidemiology of viral hepatitis A'. *Am. J. pub. Hlth*, **43**, 1097–100.

19 MCCOLLUM, R. W. (1962) 'Epidemiologic patterns of viral hepatitis'. *Am. J. Med.*, **32**, 657–64.

20 BOTHWELL, P. W. *et al.* (1963) 'Infectious hepatitis in Bristol 1959–1962'. *Brit. med. J.*, ii, 1613–16.

21 BURNS, C. (1965) 'Infectious hepatitis – its incidence in Leicester and general epidemiology'. *Roy. Soc. Hlth J.*, **85**, 144–8.

22 BURNS, C. (1967) 'Infectious hepatitis in Leicester 1963–1966'. *Br. med. J.*, iii, 773–6.

23 MILOJCIC, B. (1965) 'The influence of social factors on the spread of infectious hepatitis'. *J. Hyg. Epidem., Microbiol. Immun.*, **9**, 121–6.

24 CLARK, W. H. *et al.* (1960) 'A special study of infectious hepatitis in the general population of three counties in California'. *Am. J. trop. Med. Hyg.*, **9**, 639–51.

25 HAGGETT, P. (1965) *Locational Analysis in Human Geography*. London.

26 HAGERSTRAND, T. (1967) *Innovation Diffusion as a Spatial Process*. Chicago.

27 HAGERSTRAND, T. (1967) 'On Monte Carlo simulation of diffusion'. *Quantitative Geography*, Part I. Northwestern University, Studies in Geography Series No. 13, 1–32.

28 BARTLETT, M. S. (1956) 'Deterministic and stochastic models for recurrent epidemics'. *Proc. Third Berkeley Symposium on Mathematical Statistics and Probability*, 4, 81–109.

29 BAILEY, N. T. J. (1957) *The Mathematical Theory of Epidemics*. London.

21 Lee Wave Hypothesis for the Initial Pattern of Spread during the 1967–8 Foot and Mouth Epizootic*

R. R. Tinline

The extensive and explosive initial spread of the 1967–8 foot and mouth disease epizootic in the English Midlands has been a subject of much speculation.[1-4] After the first outbreak was confirmed at Bryn farm on October 25 (day 1), some fifty-nine further outbreaks were confirmed between days 4 and 10 in a fan-shaped area stretching north-east towards Chester, a distance of 56 km. (see Fig. 21.1). The Ministry of Agriculture has argued that, because no human or mechanical links could be established between Bryn farm and subsequent outbreaks, many of these outbreaks were probably caused by the imported frozen lamb that was thought to be responsible for the Bryn farm outbreak.[1] The only evidence for the view of the Ministry was that the lamb had been distributed near some of the outbreaks, and because the lamb had also been distributed extensively in other parts of England and Wales since August it seems very unlikely that many outbreaks should suddenly occur within a specific area and that they should occur approximately one incubation period after the Bryn Farm outbreak.

Fig. 21.1 illustrates an alternative argument that Bryn farm (or Bryn farm together with neighbouring farms that might have been emitting virus at the same time) could have been the sole source and that most of the subsequent outbreaks were the result of airborne spread from Bryn farm. While the possibility of airborne spread has been confirmed for other periods during this epizootic and for other epizootics,[2-4] my attempts to model the spread of the disease[4] have shown that the magnitude of the initial long-range spread was far greater than in any other instance of long-range spread during the remainder of the epizootic. Hence, if airborne spread was responsible for initial spread in the 1967–8 epizootic, then it was operating in a dangerously efficient manner.

Subsequent investigation revealed three curious features of the spatial pattern of initial outbreaks: the clustering of the outbreaks into three distinct clusters (see Fig. 21.1); the regular spacing of these clusters at

* Reprinted from *Nature*, (1970) **227**, 860–2.

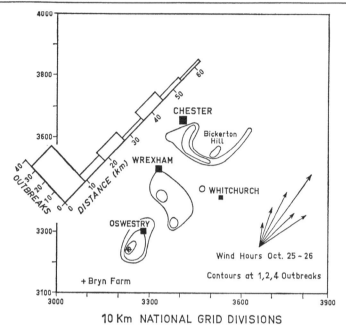

FIGURE 21.1 1967–8 Foot and mouth epizootic; outbreak for days 1–10.

18–20 km. intervals; and the fact that outbreaks in the downwind clusters occurred within the time of one incubation period after the Bryn farm outbreak. These features suggested that an atmospheric phenomenon known as lee waves may have helped the initial spread. Here I attempt to illustrate how this could have happened.

Bryn farm is located at the base of a valley ridge beside the Cheshire Plain. From October 15 to 26 the wind blew from the southwest at approximately right angles to this ridge. Fig. 21.2, a profile drawn through Bryn farm at an azimuth of 045°, makes these relationships clear. Given this situation, then in certain upper air conditions air flowing over a ridge or hill will be forced into vertical oscillation with a downwind configuration

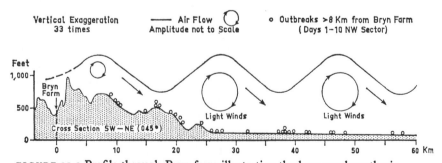

FIGURE 21.2 Profile through Bryn farm illustrating the lee wave hypothesis.

of the streamlines like that shown in Fig. 21.2. This is the phenomenon of lee waves.

It is known that the formation of lee waves can influence surface winds below the wave troughs[5] and that material can be carried aloft and deposited downwind under the influence of the waves.[6] In a similar way lee waves could have transported virus particles from the Bryn farm area out over the Cheshire Plain. A trough over the source of virus, by enhancing up-currents, may have formed an unusually effective mechanism for carrying virus particles aloft. Once aloft, each 'air parcel is carried along' the streamline pattern. Reports by glider pilots emphasise that flow in lee waves is often remarkably smooth,[7] so it seems possible that an air parcel will remain intact with a volumetric concentration of virus similar to the concentration at source. A particular air parcel might then be brought to ground level at the next trough and its virus content might be deposited or inhaled in the area immediately downwind of the trough where the winds become lighter. The remainder of the air parcel might then be carried up and on to the next trough. Such a hypothesis is especially attractive (it explains how virus from a single source could cause infections at a long range downwind), but if it is to be proven it must be shown that lee waves did occur on October 25 and/or October 26; that if they did occur they had a wavelength of approximately 18–20 km., the spacing of the clusters in Fig. 21.1; and that the crests and troughs occurred in the positions suggested by Fig. 21.2 so that pickup at source and downwind deposition could be explained.

Unfortunately, it seems that no meteorological observations were made in this area at the time, so the lee wave hypothesis cannot be directly proved or disproved. Theoretical grounds have therefore been used to estimate the occurrence and essential characteristics of these lee waves.

Table 21.1 ranks the more important of the qualitative conditions associated with the formation of lee waves[8] and indicates that October 25 and 26 were likely days for lee waves to form. Scorer[9] has shown that, with the assumptions of frictionless, steady, laminar and isentropic flow over small ridges, waves are only possible when the variable l^2 reaches a marked maximum at lower or middle altitudes and then declines rapidly with height. l^2 can be estimated from upper air soundings of temperature and wind speed. Soundings at Camborne (West Cornwall) were used in order to make sure the sample was that of the undisturbed flow over Bryn farm. Plots of temperature, wind speed and l^2 for 2330 h GMT on October 25 are given in Fig. 21.3. They show the classic symptoms of conditions favourable for the development of lee waves: an isothermal layer, wind speed increasing with height and a pronounced maximum in l^2 at middle altitudes.[9] Similar plots for 1130 h GMT on October 25 were also obtained. SIGMETS from the Flight Information Centre at Preston also predicted the occurrence of lee waves from 1630 h GMT on October 25 to 0430 h GMT on October 26 (personal communication from S. A. Casswell).

The wavelength of the lee wave formations were calculated from mean tropospheric wind speed using Corby's regression equation, which was derived from radiosonde data[8] and has been confirmed by subsequent investigators.[10, 11] The list of wavelengths in Table 21.2 shows that lee waves on the afternoon and evening of October 25 would have had the required wavelength of 18–20 km.

TABLE 21.1 *Conditions Suitable for Lee Waves*

	Camborne upper air measurements							
October:	24		25		26		27	
Condition	1130	2330	1130	2330	1130	2330	1130	2330
1 Marked stability at lower and/or middle altitudes – indicated by isothermal layer or temperature inversion	+	+	+	+	+	+	/	
2 Wind speed increasing rapidly with height	–	–	+	+	+	/	–	+
3 Little variability in wind direction with height	–	–	+	+	+	+	–	–
4 High wind speed at hill crest greater than 20 knots	–	–	+	+	–	–	–	–
5 Frontal system in vicinity (Oswestry)	–	–	+	+	–	–	–	–

+, means condition satisfied; /, means condition marginally satisfied; –, means condition not satisfied.

Finally, it must be shown that the positions of the crests and troughs of any existing pattern with a wavelength of about 20 km. would conform to the positions shown in Fig. 21.2. The underlying topography can influence the amplitude and phase of the wave and so had to be taken into account in determining the position of the wave. Wallington's method[12] was used to calculate the streamline displacement, and the result is given by Fig. 21.2. Fig. 21.2 therefore represents atmospheric conditions as they probably were during the afternoon and evening of October 25.

Conditions at this time thus seem to have been conducive to the development of waves with a period of 18–20 km. The underlying topography

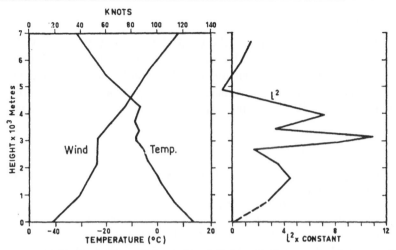

FIGURE 21.3 Temperature, wind speed and l^2 for 2330 hrs. on October 25, 1967, at Camborne.

would have ensured that the wave troughs in the air-stream formed over Bryn farm and over the downwind areas previously identified as 'clusters'. (The pattern of outbreaks in Fig. 21.2 was tested statistically with the method of Pearson.[13] The null hypothesis of randomness was rejected in favour of clustering at the 0·001 level.)

It seems that the lee wave hypothesis provides a plausible explanation of uptake of virus from a single (or localised) source, subsequent transmission at high concentration over long distances and deposition in relatively small areas downwind. Because neither primary outbreaks of foot and mouth disease nor lee waves are common events, the probability of a simultaneous occurrence of an initial outbreak at a site conducive to the development of lee waves and the occurrence of favourable conditions for lee waves seems quite low. It therefore seems that the probability of another similarly generated epizootic must also be quite low.

TABLE 21.2 *Calculated Wavelengths for October 25-26*

Oct. 25	0530 h	11·1 km.
	1130	17·5
	1730	20·4
	2330	19·4
Oct. 26	0530	16·7
	1130	11·7
	1730	13·6
	2330	9·5

ACKNOWLEDGEMENTS

I thank M. E. Hugh-Jones of the Central Veterinary Laboratory, Wey-bridge; Dr. R. Sellers of the Animal Virus Research Institute, Pirbright; Dr. E. Stringer of the Department of Geography at the University of Birmingham, and various members of the Department of Geography at the University of Bristol for helpful comments on the draft of this chapter.

REFERENCES

1 *Report of the Committee of Enquiry on Foot and Mouth Disease* (1968) Part 1. HMSO: London.
2 WRIGHT, P. B. (1969) *Weather,* **24,** 204.
3 SMITH, L. P. and HUGH-JONES, M. E. (1969) *Nature,* **223,** 712.
4 TINLINE, R. R. (1969) *Biometeorology,* Part 11, 102.
5 AANENSEN, C. J. M. and SAWYER, J. S. (1963) *Nature,* **197,** 654.
6 FÖRCHTGOTT, J. (1950) *Bull. Met. Czech.,* **4,** 14.
7 ALAKA, M. A. (ed.) (1960) *The Airflow over Mountains.* Technical Note No. 34. World Meteorological Organization.
8 CORBY, G. A. (1957) *Airflow over Mountains.* Met. Rep., No. 18. HMSO: London.
9 SCORER, R. S. (1949) *Q. Jl. R. met. Soc.,* **75,** 41.
10 FRITZ, S. (1965) *J. appl. Met.,* **4,** 31.
11 COHEN, A. and DOREN, E. (1967) *J. appl. Met.* **6,** 669.
12 WALLINGTON, C. E. (1958) *Q. Jl. R. met. Soc.* **84,** 428.
13 PEARSON, E. S. (1963) *Biometrika,* **50,** 315.

22 Contagious Processes in a Planar Graph: An Epidemiological Application

P. Haggett

The extension of the ideas of spatial diffusion from the continuous plane to a graph composed of vertices (nodes) and edges (links) is well known.[1] Recent work in epidemiology has suggested that the basic ideas of graph theory[2] may be usefully extended to the study of the spread of infectious diseases.[3] Existing applications have, however, largely been confined to epidemics in very small communities – within the household, the school, the local community[4] – or to the rather constrained spatial situations where mathematical solutions prove tractable.[5] This paper explores the application of the notions of graph theory at an intermediate spatial level to a system of contiguous geographical regions. It suggests that the operation of contagious epidemiological processes can be inferred at this gross level, but that severe data problems would need to be overcome before the concepts of graph theory could be used in any epidemic forecasting context.

DESIGN OF THE STUDY

The Study Area

The study area selected for this test was South West England. This may be conventionally defined as the six geographical counties of Cornwall, Devon, Dorset, Gloucestershire, Somerset and Wiltshire. It covers an area of 9,000 square miles, slightly larger than that of Massachusetts and in 1970 had an estimated civil population of around 3·7 m. The two main cities, Bristol and Plymouth, are the largest population concentrations. Together with their adjacent urbanised areas their estimated populations of around half a million and a quarter of a million respectively make up about one-fifth of the South West region's population. The only other major concentrations are the Gloucester–Cheltenham area (about 200,000), the Poole and Swindon areas (both about 120,000), and the Torbay area (about 100,000). There are two towns (Exeter and Bath) between 50,000 and 100,000 and a further nine towns between 20,000 and 50,000. These seventeen population clusters together make up one-half of the region's

L

population; the remainder is spread more sparsely through a system of smaller towns and rural settlements.

For statistical purposes the main units used by the General Register Office (GRO) are the local authority administrative areas: the county boroughs (C.B.), municipal boroughs (M.B.), urban districts (U.D.) and rural districts (R.D.). The South West is divided into 179 such areas. If we ignore the Scilly Isles, the remaining 178 areas constitute a contiguous network of areas that may be represented in the form of an unweighted planar graph. Each GRO area represents a vertex on such a graph, while the existence of a common boundary between any pair of areas is indicated by a link. Other and more complex forms of link definition and weighting are discussed below (see Table 22·4), but even the simple binary method adopted allows some definition of the area's characteristics in graph terms. Areas lying on the border of the South West region at its junction with other regions (the geographical counties of Monmouthshire, Hereford, Worcestershire, Warwickshire, Oxfordshire, Berkshire and Hampshire) have the number of their link contacts arbitrarily reduced by the boundary effect, i.e. they have links to contiguous GRO areas which because they lay outside the boundaries of the South West were not studied. A 'buffer zone' was therefore distinguished to avoid this undercounting effect.

The remaining graph consists of 151 vertices with an average of 3·22 links per vertex. The difference between this and the expected value of 5·79 links per vertex for a random set of contiguous areas[6] reflects two peculiarities of the South West region. First, its peninsular form with a large number of areas abutting on its land–sea boundary. Second, the large number of urban areas (C.B.'s, M.B.'s and U.D.'s) that are completely surrounded by single rural areas (R.D.'s) and then are recorded as having a single link.

Analysis of the GRO area system as a graph allows its direct comparison with other regions. The South West system is marked by its long diameter (twenty-two links) in relation to the average link distance between vertices (seven links). The degree of connectivity, 249 links as against a possible maximum of 447 links for a planar graph with its observed 151 nodes. Again the relative location of individual vertices with respect to others in the network can be specified (Table 22.1).

Epidemiological Data

For England and Wales a major published source of epidemiological data is the Registrar General's *Weekly Return*. This includes notifications of certain infectious diseases as supplied by the Medical Officer of health for each GRO area for the week (ending on a Friday). Detailed clinical work suggests that notifications may seriously underestimate the actual incidence of a disease. Apart from variations in diagnosis, the practice of individual medical practitioners in notifying cases to local Medical Officers of Health is known to vary widely. Not enough information is available about regional variations to allow application of individual correction factors but these might be expected to run around $\times 1 \cdot 5$ to $\times 2 \cdot 00$ for the notifications actually recorded.

The selection of measles notifications from the group of infectious diseases for which data was available was determined by a number of

TABLE 22.1 *Relative Location of Three Major Outbreak Centres in Relation to South West Region*

GRO Area:	Bristol C.B.	Plymouth C.B.	Penzance U.D.
Total link distance	1,090	1,141	2,576
Mean link distance	6·12	8·02	14·47
Maximum link distance	16	15	22
% GRO areas within three links	24%	11%	6%

factors. First, the high rate of incidence yielded a large number of notifications (over 250 a week for the whole South West region over the period studied). Second, the transmission of the disease from person to person without the presence of an intermediate host allowed demographic data to be combined directly in the analysis. Third, the highly contagious nature of the disease suggested that distance-decay factors would operate somewhat strongly in guiding the spatial pattern of outbreaks. Finally, measles has played a central role in the development of quantitative epidemiological theory[5, 7] and a number of the classic deterministic and stochastic models of epidemic spread were first derived from consideration of measles returns.

SPATIAL PATTERNS OF OUTBREAKS

For the South West region weekly returns were studied over a sixty-week period between September 1969 and October 1970 (Fig. 22.1). For this period 16,496 cases of measles were notified spread over 10,740 recording units (179 areas × 60 weeks); an indicated average of 1·54 notifications per area per week. In practice, however, the notifications were strongly clustered in both time and space, so that over 70 per cent of the basic recording units showed zero returns. The average over the units actually reporting cases is therefore considerably greater (3·78 notifications per area per week).

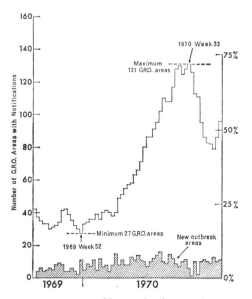

FIGURE 22.1 Change in the number of GRO areas within the South West region with measles notifications over the study period, weeks 1969/36 to 1970/43 inclusive.

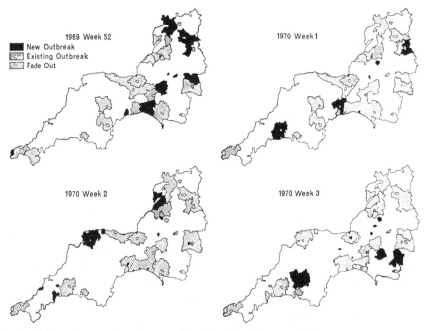

FIGURE 22.2 Changing distribution of measles notifications in GRO areas in the South West region in weeks 1969/52 to 1970/3 inclusive.

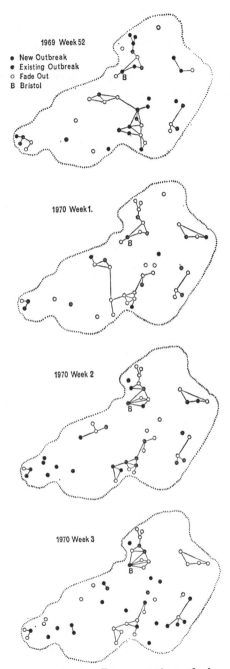

FIGURE 22.3 Representation of the same information as in Figure 22.2 in a graph format. Adjacent GRO areas are indicated by a link in the graph.

Mapping of each week's returns on a choropleth map (Fig. 22.2) and on a graph (Fig. 22.3) allowed an atlas of the spatial spread to be built up. This showed that epidemics were recurrent over the region with a distinctive 'waxing and waning' or 'swash and backwash' pattern based on five major centres, viz. greater Bristol, Plymouth, west Cornwall, south east Dorset and south Wiltshire, and the Gloucester–Cheltenham area. At the lowest point of the sixty-week period only twenty-seven GRO areas had outbreaks (about one area in seven), while at the maximum over 130 were infected (about two areas in three).

The Location of 'New' Outbreaks

The first topic investigated was that of the location of new outbreaks. What are their spatial characteristics? Where are they located in respect of existing outbreaks? The definition of an outbreak in any GRO area as used in the paper is based on the recorded notifications in the *Weekly Returns* together with an assumed 'fade out' period of two consecutive weeks.[8] An outbreak is assumed to have ended when a GRO area fails to report new cases for three or more consecutive weeks and to begin again with the next reported case (i.e. one or more notifications in any week).

The location of 'new' outbreaks in respect to existing outbreaks is of special interest if we assume (i) that such outbreaks represent the reinfection of temporarily free areas from source areas (either inside or outside the South West) where the epidemic is being maintained and (ii) that *ceteris paribus* the probability of population mixing is higher for adjacent areas than for very distant areas. Some of the other considerations affecting mixing rates are discussed below (p. 320). It should be stressed that the definition of 'new' outbreaks used here is entirely dependent on the GRO notifications, and that, because of under-reporting, outbreaks may be erroneously defined as new which in reality are the continuation of an existing outbreak.

As defined above, the 179 South West areas recorded some 532 outbreaks over the sixty-week period, a gross average of 2·97 outbreaks per area. However, by examining the returns for a 'lead in' period for the fortnight preceding the first week of the study period it was possible to establish that forty of the GRO areas were recording outbreak conditions at the beginning of the period. This left 492 new outbreaks available for study. Table 22.2 summarises the characteristics of these outbreaks. Half the outbreaks were less than four weeks in duration and involved four or less notifications. Fig. 22.4 shows the distribution of the outbreaks in time, their duration, and the size relationships of the largest outbreaks.

The computed values of the contact probabilities are shown in Table 22·3. Of the 4,936 GRO areas examined, some 52·8 per cent were contiguous

TABLE 22.2 *Characteristics of Recorded Outbreaks in South West Region*

	Total	Outbreak Characteristics			
		Arithmetic Mean	*Median*	*Interquartile Range*	*Weighted Mean**
Notifications	16,496	31·01	4	1–35	35·02
Outbreak weeks	4,369	8·21	4	3–10	9·28

Source: Office of Population Censuses and Surveys. Registrar General's Weekly Return for England and Wales. Weeks 1969/36 to 1970/43 inclusive.

* Outbreaks truncated by the boundaries of the sixty-week study period counted as 0·5 (one boundary) or 0·25 (both boundaries).

with areas already infected or in a 'fade out' phase. With increasing link distance from the outbreak areas the proportion of areas fell, until at four links only 1·1 per cent of areas were recorded. No areas were recorded at five links or more than five links. The pattern of new outbreaks showed a distinctly more contagious distribution with three-quarters of the 416 new outbreaks occurring in directly contiguous areas. By three links distant the

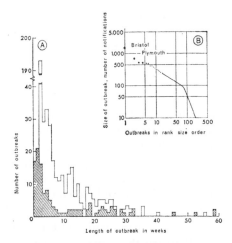

FIGURE 22.4 Characteristics of measles outbreaks occurring in the South West region, Weeks 1969/36 to 1970/43 inclusive. (A) Duration of individual outbreaks in weeks. Shaded areas indicate truncated outbreaks extending outside the sixty-week study period. (B) Size distribution of the 160 largest individual outbreaks with ten or more notifications.

proportion had fallen to less than 5 per cent and no new outbreaks were recorded four or more links from an existing outbreak.

The declining incidence of new outbreaks with distance from existing outbreaks is illustrated by a comparison of the probabilities shown in the final column of Table 22.3. The overall probability of an uninfected area having a new outbreak over the study period was about 1/12. For the areas in contact with an existing outbreak the probability is substantially greater, 1/8. At three links distance the probability has fallen to around 1/31.

TABLE 22.3 *Changing Probabilities of New Outbreaks with Contact Distance from Existing Outbreak**

Contact Position	GRO areas outside existing outbreaks		New outbreaks		Probability of new outbreak	
	Number	%	Number	%		
One link	2,603	52·74	316	75·96	·1214	1/8
Two links	1,749	35·43	83	19·95	·0475	1/21
Three links	529	10·72	17	4·09	·0321	1/31
Four links	55	1·11	–	–	–	
TOTAL	4,936	100·00	416	100·00	·0843	1/12

* South West Region Weeks 1969/36 to 1970/43 inclusive.

The Location of Persistent Outbreaks

The second topic investigated was the location of persistent outbreaks. What characterised areas with a persistent pattern of notifications? For the South West region as a whole the number of new cases reported over the sixty-week period averaged 560 per fortnight, and although the number of new cases varied widely over that period, at no point was the maintenance of the infection in doubt. For the 179 individual GRO areas, however, only the two with the largest populations (Bristol C.B. and Plymouth C.B.) showed an unbroken record; conversely nine of the areas recorded no new cases over the period. Between these two extremes stand a sequence of intermediate cases with highly variable patterns of notification (see Fig. 22.5).

The fortnightly average is of importance in relation to the incubation period of the disease. Measles is characterised by the epidemiologist as a highly infectious virus disease with an incubation period of about ten days to fourteen days depending on the symptoms described. Generalizing, one can say that the length of time between cases in any chain of infection is

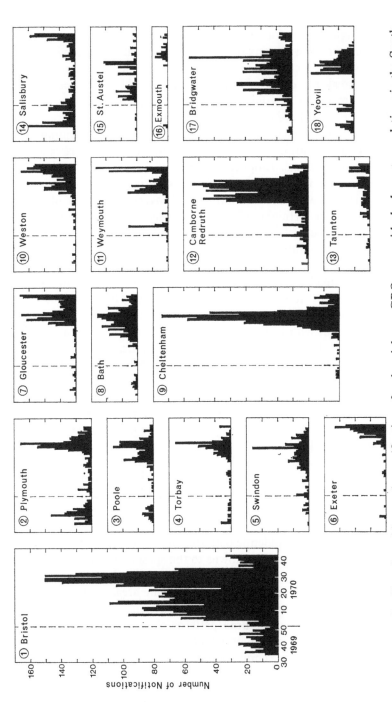

FIGURE 22.5 Profiles of weekly measles notifications for the eighteen GRO areas with the largest populations in the South West region. In each case the profiles are for aggregated GRO areas in which the notifications have been combined with those of contiguous areas.

about a fortnight. Therefore to maintain measles in a community an absolute minimum of twenty-six new cases a year would be needed to keep an epidemic going. Since there is a high probability that some factors will intervene to break the chain of infections, we should need considerably more cases to maintain an epidemic. As Bartlett[9] points out, this phenomenon 'suggests that there will be a critical community size, above which measles should tend to maintain itself, whereas for smaller communities it will die out and require re-introducing from outside before another epidemic can materialize'. Empirical studies of weekly measles notifications for various towns in England and Wales during the period 1940–56 confirmed this view with the critical threshold covering the range of 114 to 134 notifications per fortnight (i.e. Hull and Bristol respectively). Towns with fewer notifications showed intervals between epidemics with the size of the interval consistently related to the inverse of total community size.[8]

Parallel work[10] on North American cities over the twenty-year period 1921–40 was limited by the monthly (rather than weekly) reporting period but again suggested the existence of a critical level. Here the four most critical cities were Akron, Providence, Rochester (United States) and Winnipeg (Canada) with average fortnightly notifications of around 162, or about half the England and Wales figure. Application of correction factors to allow for incomplete notification brought up the England and Wales figure to 212 and the North American figure to 160 cases per fortnight.

The sixty-week period studied in the South West is too short to allow replication of Bartlett's tests. The only two areas to maintain a continuous infection were Bristol C.B. with a recorded average of forty-seven notifications per fortnight and Plymouth C.B. with fourteen. The next two areas with substantially lower populations maintained a continuous outbreak for all but four weeks of the study period with values of seventeen (Sodbury R.D. contiguous with Bristol C.B.) and ten (Bridgwater R.D.C.) (Fig 22.5).

Gradual change in the epidemic pattern from the continuous succession of peaks and troughs of the large towns to the spasmodic outbreaks of the small community is well known. However, examination of the weekly returns for different towns show that while some areas remain in phase with the large towns, others lead or lag in a characteristic way. A full understanding of the kind of space-time patterns shown in Fig. 22.5 would involve a detailed tracing of individual infections or a very comprehensive mathematical model or some combination of the two. The conventional use of cross-spectral analysis to examine the coherence and phase lags between pairs of such series[11] and the development of forecasting models[12] is limited by the discontinuous nature of the series. However, recent work on the spectrum of 'clipped' series[13] promises to overcome this objection.

DISCUSSION

The crude results reported here clearly represent only a first approximation to the accurate estimation of epidemic spread through a graph. Apart from the self-evident need to check the results obtained over a longer time period and for other regions, a number of major modifications in the definition of the graph itself seem desirable.

Re-definition of Graph Vertices

The use of the existing pattern of GRO areas, to represent the vertices of the graph, is open to criticism and substitution of aggregated areas, to give geographical units more closely in line with the socio-economic realities of population distribution, is indicated. Most county boroughs, municipal boroughs and urban districts are 'underbounded' in that the built-up urban area extends beyond the present administrative boundaries into the surrounding rural districts. Information for journey-to-work patterns from the 1966 Sample Census confirms the extent of the daily interchange of population across the GRO boundaries (see Table 22.4). Interestingly the revised units suggested by the Royal Commission on Local Government in England and Wales (the Redcliffe–Maud Commission) would generally reduce the artificial town–country distinction of the present areas and would yield improved units for epidemiological reporting. Use of such aggregated units would permit more accurate estimation of the thresholds at which contagious diseases remained persistent. Comparison of the results for the single GRO areas with those obtained by averaging of the area plus its neighbours show considerable contrasts: compare, for example, the graphs of Plymouth and Yeovil in Fig. 22.6.

In giving equal weight to all GRO areas the study reported here failed to reflect the immense differences in the size and the demographic structure of the individual units. Within the South West systems the dominant role is played by Bristol C.B. which in the 1966 census accounted for 12 per cent of the whole region's population. At the other extreme the smallest R.D. had a population some orders of magnitude smaller with only a few thousand inhabitants. Between these two the population was arrayed in approximately an inverse rank-size order.

To size differences are added substantial differences in demographic structure. Rural districts bordering large urban centres commonly have a high proportion of suburban overspill with young families with a high proportion of primary-school age children. Thornbury R.D. and Highworth R.D. are cases in point. At the other extreme, some south coast towns play an important role as retirement areas and have a relatively high proportion of population aged 60 years and over. (e.g. Torbay C.B. and Budleigh Salterton U.D.). Areas should clearly be weighted in relation to

TABLE 22.4 *Alternative Weighting of Links between Bristol C.B. and Contiguous GRO Areas*

	Estimated Workplace –Residence Flows	Link Weightings	
		Flow magnitude	Flow in relation to local GRO populations
CONTACT ZONE I			
Sodbury R.D.	20,590	1·000	·383
Kingswood U.D.	8,570	·416	·570
Long Ashton R.D.	6,830	·332	·611
Thornbury R.D.	6,280	·305	·365
Mangotsfield U.D.	5,520	·268	·868
Keynsham U.D.	4,570	·222	·435
Warmley R.D.	4,430	·215	1·000
Bathavon R.D.	1,070	·052	·178
Zone I Average	7,232	·351	·551
CONTACT ZONE II			
Bath C.B.	1,740	·085	·031
Axbridge R.D.	1,630	·079	·101
Clutton R.D.	1,580	·077	·197
Portishead U.D.	1,210	·059	·265
Clevedon U.D.	1,190	·058	·212
Calne-Chippenham R.D.	<260	<·013	<·016
Norton Radstock U.D.	130	·006	·015
Dursley R.D.	<120	<·006	<·009
Four GRO areas with flows of less than fifty persons	<400	<·019	<·003
Zone II Average	<690	<·034	<·071
CONTACT ZONE III			
Weston-super-Mare M.B.	1,210	·059	·054
Gloucester R.D.	<130	<·006	<·004
Chippenham M.B.	<100	<·005	<·007
Trowbridge U.D.	<100	<·005	<·007
Bridgwater R.D.	<100	<·005	<·012
Seventeen GRO areas with flows of less than fifty persons	<1,170	<·083	<·025
Zone III Average	<152	<·007	<·005
Other Parts of Great Britain	>2,480	–	–

Source: GRO Sample Census 1966 England and Wales. Workplace and Transport Tables, Part I, Tables 1, 2 and 4. County Reports, Gloucestershire, Somerset, Wiltshire.

the age structure in an attempt to estimate the 'at risk' population. However since the last detailed picture of age structure is only available from the 1966 Sample Census, procedures such as those suggested by Myers[14] would be needed to bring the figures up to the 1969–70 level.

Re-definition of Graph Links

The measurement of linkage between GRO areas was here determined solely by the presence or absence of a common boundary. It will be clear from inspection of the map that this simple criteria of geographical proximity is a crude one and that it would be possible to move from this simple binary scale to finer yardsticks using measures such as the length

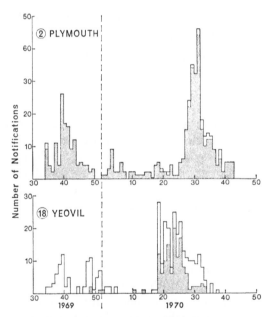

FIGURE 22.6 Contrasting effect of aggregation of adjacent GRO areas on the pattern of measles notification. The record for Plymouth C.B. is altered very little; that of Yeovil M.B. is considerably raised.

of common boundary. A more important modification, however, concerns the nature of the contact graph itself. As we have already noted, the graph of the GRO network is planar with only 249 links connecting the 151 vertices. It is clear that modern transport allows infections to occur between GRO areas *without* intervening areas being placed at risk. Theoretically therefore each of the areas may be regarded as being linked to all of the others; with the vertices in the South West region the total number of such links would be 11,325. However, while all such links are possible the evidence of Table 22.3 suggests that local links may be predominantly important in terms of the spread. One possible way of estimating the importance of all alternative links would be to use a 'gravity' type model of the form:

$$W_{ij} = \log \beta_0 + \beta_1 \log P_i + \beta_2 \log P_j - \beta_3 \log D_{ij}$$

where

W_{ij} = the weight to be assigned to a given link IJ in the GRO network

P_i, P_j = estimated size of the 'susceptible' population in GRO areas I and J respectively

D_{ij} = a measure of the intervening distance between GRO areas I and J

β_0, etc. = empirically determined constants

Calibrating such a model would itself give rise to a number of acute problems. Use of available area-interchange data on journey-to-work (Table 22.4) would clearly overestimate the spread effect and give far too low values for the critical β_3 constant. Estimates of movements from journey-to-school data might yield more acceptable values. Although systematic data is not available on this point, the general form of the distance-decay curves are known.

Substitution of Alternative Graphs

From the point of view of measles *per se* the GRO areas have substantial merit. Measles is most prevalent in young children and most movements of primary school children (approximate ages five through eleven years) between home and school take place *within* the local authority areas. Crossing of boundaries is more important for primary-school children attending either denominational schools or private schools outside the state school system (less than 10 per cent of the total). At age eleven, the degree of inter-area 'mixing' of school populations increases since most secondary-school children (approximately ages eleven through eighteen) with homes in the rural districts attend schools located in the urban districts.

Although no regularly published epidemiological information is available for individual schools, study of school attendance registers produce data of some interest. Fig. 22.7 shows profiles of recorded absence rates for two schools in Somerset over four winter periods. Rates vary from school to school and show considerable variation on the national average rate of 7·2 per cent loss of schooling (for children between $6\frac{1}{2}$ and $10\frac{1}{2}$).[15] Special interest attaches to the sharp peaks when the loss soars above the 50 per cent level. The problems in screening out from such peaks non-epidemiological causes (e.g. heavy snowfall or the local fair!) and systematically matching the remaining peaks against local MOH records remains formidable and lie outside this paper's immediate concern. However, as

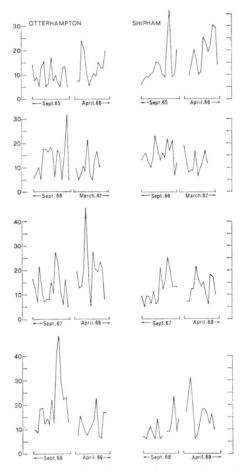

FIGURE 22.7 Profiles of percentage loss of schooling over four winter sessions (September to March/April) for children aged between five and seven years for two sample schools, Otterhampton (Bridgwater R.D.) and Shipham (Axbridge R.D.). Data from registers of attendances maintained at the schools.

Fig. 22.8 shows, the geographical pattern of schools is so much more close than that of areas (approximately 2,700 schools as against 179 GRO areas for the South West) and the possibility of tracing epidemics (e.g. influenza) at this more detailed level so attractive that more work seems worthwhile.

Where local education authorities demarcate 'collecting areas' around each school, these may be converted to graphs in the conventional way. Failing this Dirichlet regions could be set up around each school location,[16] a graph constructed directly from a modified nearest-neighbour routine, or census tracts aggregated to form reasonable hinterland units.

The accurate forecasting of the spatial spread of a disease remains one of the central goals of epidemiology and its utility in determining a disease-control strategy is self-evident. To date, the main advances have come from (i) detailed empirical studies of disease transmission within families and small communities by practical epidemiologists and (ii) broad mathematical

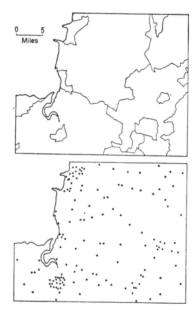

FIGURE 22.8 Contrasts between the geographic scale of the recording grid for (A) GRO areas and (B) schools for the north Somerset section of the South West region.

and statistical models based on deterministic or stochastic diffusion models. Attempts to combine the two approaches by devising models, which are both general in application but involve reasonable epidemiological assumptions, have run into some formidable mathematical problems not all of which have been solved. Treatment of sets of discrete areas as graphs may allow some limited advances in this process of bridge building. Estimation of the probability of spread from one area to those directly (or indirectly) in contact with it may supply one element, albeit a very minor one, in assembling a more general epidemiological model of recurrent epidemics.

CONCLUSION

This paper has reported some crude attempts to use the concept of a set of contiguous regions as a planar graph in the analysis of epidemiological data. The results obtained confirm that a distinct spatial-spread element is observable when the location of new outbreaks is treated in graph terms. A number of modifications and improvements are suggested. It should be stressed that such results are limited not only by the particular geographical characteristics of the area studied but more particularly by the incomplete and variable nature of the original data.

ACKNOWLEDGEMENTS

I am grateful to Mrs. Linda Campbell for help in preparing and analysing the statistical returns.

REFERENCES

1 HAGGETT, P. and CHORLEY, R. J. (1969) *Network Analysis in Geography.* London.
2 FLAMENT, C. (1963) *Application of Graph Theory to Group Structure.* Englewood Cliffs, N.J.
3 TAUTU, P. (1970) 'Structural models in epidemiology: an introductory investigation'. World Health Organization: Seminar on Quantitative Epidemiology. Moscow, November 23–7 1970. Unpublished paper.
4 ELVEBACK, L., FOX, J. P. and VARMA, A. (1962) 'An extension of the Reed-Frost epidemic model for the study of competition between viral agents in the presence of interference'. *Am. J. Hyg.*, **80**, 356–64.
5 BAILEY, N. T. J. (1955) 'Some problems in the statistical analysis of epidemic data'. *Jl. R. Statist. Soc.*, Series B. **17**, 35–68.
  BAILEY, N. T. J. (1957) *The Mathematical Theory of Epidemics.* London.
  PIKE, M. C. and SMITH, P. G. (1968) 'Disease clustering: a generalization of Knox's approach to the detection of time-space interactions'. *Biometrics*, **24**, 541–56.
6 HAGGETT, P. and CHORLEY, R. J. (1969) *op. cit.*, pp. 50.
7 STOCKS, P. and KARN, M. (1928) A study of the epidemiology of measles. *Annals of Eugenics*, **3**, 361–98.
  SOPER, H. E. (1929) 'The interpretation of periodicity in disease prevalence'. *Jl. R. Statist. Soc.*, **92**, 34–61.

DIETZ, K. (1967) 'Epidemics and rumours: a survey'. *Jl. R. Statist. Soc.*, Series A, **130**, 505–28.

8 BARTLETT, M. S. (1957a) 'Measles periodicity and community size, *Jl. R. Statist. Soc.*, Series A, **120**, 48–70.

9 BARTLETT, M. S. (1960) *Stochastic Population Models in Ecology and Epidemiology*, pp. 66. London.

10 BARTLETT, M. S. (1957b) 'The critical community size for measles in the United States'. *Jl. R. Statist. Soc.*, Series A, **123**, 37–44.

11 BASSETT, K. and HAGGETT, P. (1971) 'Towards short-term forecasting for cyclic behaviour in a regional system of cities'. In: M. D. I. Chisholm, A. E. Frey, and P. Haggett (Eds) *Regional Forecasting*. London.

12 BOX, G. E., JENKINS, G. M. and BACON, D. W. (1967) 'Models for forecasting seasonal and non-seasonal time series. In: B. HARNS (Ed.) *Advanced Seminar on Spectral Analysis of Time Series*. New York.

13 VAN VLECK, J. H. and MIDDLETON, D. (1966) 'The spectrum of clipped noise'. *Proc. Instn. elec. electron. Engrs.*, **54**, (i), 2–19.

14 MYERS, C. L. (1967) Forecasting of child populations. Local Government Operational Research Unit Report.

15 KAHN, J. H. and NURSTEN, J. P. (1965) *Unwillingly to School*. Oxford.

16 HAGGETT, P. and CHORLEY, R. J. (1969) *op. cit.*, pp. 236.

Index